AF588081

SEMA SIMAI Springer Series

ICIAM 2019 SEMA SIMAI Springer Series

Volume 12

This sub-series of the SEMA SIMAI Springer Series aims to publish some of the most relevant results presented at the ICIAM 2019 conference held in Valencia in July 2019.

The sub-series is managed by an independent Editorial Board, and will include peer-reviewed content only, including the Invited Speakers volume as well as books resulting from mini-symposia and collateral workshops.

The series is aimed at providing useful reference material to academic and researchers at an international level.

More information about this subseries at http://www.springer.com/series/16499

Gabriella Puppo • Andrea Tosin
Editors

Mathematical Descriptions of Traffic Flow: Micro, Macro and Kinetic Models

Springer

Editors
Gabriella Puppo
Department of Mathematics
Sapienza University of Rome
Roma, Italy

Andrea Tosin
Department of Mathematical Sciences
"G. L. Lagrange"
Politecnico di Torino
Torino, Italy

ISSN 2199-3041 ISSN 2199-305X (electronic)
SEMA SIMAI Springer Series
ISSN 2662-7183 ISSN 2662-7191 (electronic)
ICIAM 2019 SEMA SIMAI Springer Series
ISBN 978-3-030-66559-3 ISBN 978-3-030-66560-9 (eBook)
https://doi.org/10.1007/978-3-030-66560-9

This Springer imprint is published by the registered company Springer Nature Switzerland AG.
The registered company address is: Gewerbestrasse 11, 6330 Cham, Switzerland

Preface

Mathematical models for traffic flow are a tool to predict and possibly control the behaviour of vehicular traffic. These models can be at the basis of policy decisions with the aim of improving the fluidity of traffic, such as for instance cross lights temporization, yield rules at roundabouts and one-way road planning in urban areas. Future developments, such as the introduction of automatically driven cars, will require new policy decisions with the aim of protecting drivers from accidents while maximizing roads capacity. The presence of reliable mathematical tools to study these phenomena can yield important contributions to decision making. But, the possibility of using mathematical modelling to address these questions will depend crucially on the availability of reliable and robust mathematical models for traffic flow. The contributions gathered in this book are a step towards the construction and the analysis of mathematical models for traffic.

There are several levels in which a mathematical description of traffic can be carried out. The simplest level is the microscopic level, where each vehicle is considered as a single particle, interacting with its neighbours. The model consists mainly in the mathematical definition of the interaction rules between the different particles, which determine the acceleration or the deceleration of the single vehicle. Next, the complete dynamics of the flow is obtained considering the system of ordinary differential equations generated by the evolution dynamics of all vehicles present in the flow. Since each car generates an equation, these models permit a detailed description of traffic flow, but, at the same time, they are computationally costly, and too detailed for the answers one is typically interested in, which concentrate on global quantities, as the propagation and formation of traffic jams.

Macroscopic models study the evolution of traffic flow as a bulk, without considering the behaviour of the single vehicles. They study the evolution of global quantities, such as the traffic density or the mean speed of the flow in a certain section of the road. Thus, macroscopic models usually contain a few partial differential equations for the evolution of the quantities of interest. As a consequence, they are computationally much faster than microscopic models, but they need ad hoc closure relations that substitute the interaction rules designed in microscopic models. And, these closure relations can have a very loose link with the physics of traffic.

Mesoscopic (or kinetic) models are a bridge between the microscopic and the macroscopic world. Here, one starts from the construction and the rules of a microscopic model and builds a distribution function that contains a statistic of all the particles crowding the microscopic world. Using the methods from mathematical physics, this distribution function is evolved in time, and its moments, i.e. its average values, provide the global quantities characteristic of macroscopic models and their evolution. In this fashion, one obtains closure rules for macroscopic models, which have deep roots in the underlying microscopic models.

This book contains a collection of papers covering recent developments in the theory of traffic flow, covering all three approaches. The first paper, "Reconstruction of Traffic Speed Distributions from Kinetic Models with Uncertainties", considers the construction of microscopic models taking into account the heterogeneous behaviour of drivers in their interactions, and the complexity due to the presence of different types of vehicles. The recent technique of uncertainty quantification is applied to take into account the possible lack of information on the details of the single interactions and the characteristics of the vehicles composing the flow. The model is recast in a statistical sense, and comparisons with experimental data confirm the quality of the proposed framework.

The second paper, "From Kinetic to Macroscopic Models and Back", considers the construction of kinetic models from microscopic ideas, concentrating on the need to account for unstable phenomena that are responsible for the onset of traffic jams, and in particular of phantom traffic jams, arising in dense traffic flow through the onset of stop-and-go waves. The paper shows how and why second-order macroscopic models with relaxation terms are able to account for the onset of stop-and-go waves, as long as the classical Chapman–Enskog analysis is extended to include regions of the flow with negative diffusion. A macroscopic second-order model is derived from a kinetic approach, stemming from individual interaction rules at the microscopic level.

The third paper in this series, "Structural Properties of the Stability of Jamitons", continues the study of phantom traffic jams—or jamitons—in second-order macroscopic models. The paper studies the amazing complexity of the solutions that can be originated even from simple traffic models, addressing the development and the evolution of jamitons, including the coalescence and fragmentation of these waves. The fourth contribution "Stop-and-Go Waves: A Microscopic and a Macroscopic Description" also concentrates on the development of unstable waves, constructing microscopic and macroscopic models starting from experimental data, instead of heuristic or modelling considerations.

Finally, "An Overview of Non-local Traffic Flow Models" studies a different approach to the construction of macroscopic models for traffic flow. Instead of allowing for complex phenomena thanks to second-order models, as in the first paper of this collection, here the shortcomings of classical first-order models, based on a single scalar equation, are addressed enriching the model with non-local terms in the flux function. The paper discusses the modelling strategy behind the introduction of non-local terms and addresses the mathematical well-posedness of the resulting models.

Roma, Italy
Torino, Italy
August 2020

Gabriella Puppo
Andrea Tosin

Contents

Editors and Contributors

About the Editors

Gabriella Puppo is professor of numerical analysis at the Sapienza, Universitá di Roma. Previously, she has had positions at the Universitá dell'Insubria and Politecnico di Torino. She obtained a Ph.D. in Applied Mathematics at the Courant Institute of NYU. Her research interests range from scientific computing for hyperbolic and kinetic equations to modelling of physical and social phenomena using kinetic theory and hyperbolic PDEs.

Andrea Tosin is full professor of Mathematical Physics at the Politecnico di Torino, Italy. His main research interests consist in revisiting the classical methods and concepts of kinetic theory, such as Boltzmann-type collisional equations and Fokker–Planck asymptotics, to investigate emerging problems in the realm of interacting multi-agent systems. Applications include vehicular traffic, crowd dynamics and social systems.

Contributors

Caterina Balzotti SBAI Department, Sapienza Università di Roma Rome, Italy

Felisia Angela Chiarello Inria Sophia Antipolis Mediterranée, Université Côte d'Azur Sophia Antipolis, France

Michael Herty RWTH Aachen University, Aachen, Germany

Elisa Iacomini SBAI Department, Sapienza Università di Roma Rome, Italy

Gabriella Puppo "La Sapienza" Università di Roma, Roma, Italy

Rabie Ramadan Department of Mathematics, Temple University, Philadelphia, PA, USA

Rodolfo Ruben Rosales Department of Mathematics, Massachusetts Institute of Technology Cambridge, MA, USA

Benjamin Seibold Department of Mathematics, Temple University, Philadelphia, PA, USA

Andrea Tosin Politecnico di Torino, Torino, Italy

Giuseppe Visconti RWTH Aachen University, Aachen, Germany

Mattia Zanella University of Pavia, Pavia, Italy

Reconstruction of Traffic Speed Distributions from Kinetic Models with Uncertainties

Michael Herty, Andrea Tosin, Giuseppe Visconti, and Mattia Zanella

Abstract In this work we investigate the ability of a kinetic approach for traffic dynamics to predict speed distributions obtained through rough data. The present approach adopts the formalism of uncertainty quantification, since reaction strengths are uncertain and linked to different types of driver behaviour or different classes of vehicles present in the flow. Therefore, the calibration of the expected speed distribution has to face the reconstruction of the distribution of the uncertainty. We adopt experimental microscopic measurements recorded on a German motorway, whose speed distribution shows a multimodal trend. The calibration is performed by extrapolating the uncertainty parameters of the kinetic distribution via a constrained optimisation approach. The results confirm the validity of the theoretical set-up.

1 Introduction

Similarly to classic rarefied gas dynamics, kinetic modelling for traffic flow needs to define basic dynamics on microscopic entities. In particular, traffic "particles" are the vehicles, which modify their speed according to some binary interaction laws, whose definition may impact on the aggregate description and on the limit hydrodynamic trends, see [6, 7, 9, 15, 23, 29]. Furthermore, when describing behavioural phenomena the physics-inspired methods of kinetic theory needs to face new challenges, since interaction forces cannot be inferred from first principles and

M. Herty (✉) · G. Visconti
RWTH Aachen University, Aachen, Germany
e-mail: herty@igpm.rwth-aachen.de; visconti@igpm.rwth-aachen.de

A. Tosin
Politecnico di Torino, Torino, Italy
e-mail: andrea.tosin@polito.it

M. Zanella
University of Pavia, Pavia, Italy
e-mail: mattia.zanella@unipv.it

G. Puppo, A. Tosin (eds.), *Mathematical Descriptions of Traffic Flow: Micro, Macro and Kinetic Models*, SEMA SIMAI Springer Series 12,
https://doi.org/10.1007/978-3-030-66560-9_1

physical forces are replaced by empirical social forces. These new interactions are typically deduced heuristically with the aim to reproduce the qualitative behaviour of the system and are at best known with the aid of statistical methods.

Once a sound kinetic model is available, its effectivity can be measured in terms of its ability to replicate and forecast system dynamics. Nevertheless, the uncertainty which is present at the level of particles may have a very strong effect at different scales. In addition, the most used methods coming from uncertainty quantification, such as generalised polynomial chaos or collocation methods, typically assume the knowledge of the uncertainty distribution in order to develop accurate solvers, see [3, 12, 31]. Unfortunately, structural uncertainties in social systems may be highly non-standard and may change in time due to external influences.

In this chapter we aim at theoretical insights into the extrapolation of the statistical distribution of the uncertainty starting from data on traffic dynamics collected within the project [14]. In particular, we will try to calibrate the speed distribution predicted by a kinetic traffic model taking advantage of the knowledge of the empirical one obtained from real data. We will show how the microscopic uncertainty is naturally transferred to the observable quantities and, based on the measured mixed traffic conditions, we will propose an approach that catches the empirical speed distributions in several traffic regimes. We think that the promising results produced by the present approach may be useful both for the prediction and for the accurate reconstruction of real phenomena after sensitivity analysis.

This chapter has the following structure: in Sect. 2 we introduce the theoretical set-up of the problem with emphasis on the role of the uncertainty in the modelling of microscopic interactions. We also introduce a Boltzmann-type kinetic approach for traffic dynamics and, in a suitable asymptotic regime, we compute the equilibrium speed distributions, which depend on the microscopic uncertainty. Moreover, we define the quantities of interest to be compared with the traffic data described in Sect. 3. Finally, in Sect. 4 we perform the calibration of the equilibrium distributions in several traffic regimes through constrained optimisation techniques.

2 Kinetic Modelling with Uncertain Interactions

Traditional microscopic traffic models are based on the assumption that the traffic stream is composed by indistinguishable vehicles, whose reaction to speed changes is linked to the vehicle type. However, structural differences between vehicles are often observed in real traffic flows in terms, for example, of vehicle weight, engine efficiency and more or less aggressive driver behaviour. The traffic heterogeneity influences the deceleration/acceleration process of drivers in mixed traffic conditions, see [17, 18]. Experimental evidence of this fact and of the relation with traffic safety issues has been recently presented in [5].

In the following we will show that, thanks to the theoretical tools provided by uncertainty quantification, we may easily describe the aggregate trends taking care of the structural heterogeneity of real traffic flows.

Let us characterise the microscopic state of two interacting vehicles by means of their dimensionless and normalised speeds $v,\ v_* \in [0,\ 1]$. We describe the post-interaction speeds $v',\ v'_*$ in terms of the following scheme

$$\begin{cases} v' = v + \gamma I(v,\ v_*;\ z) + D(v)\eta \\ v'_* = v_*. \end{cases} \tag{1}$$

In (1) the quantity $\gamma \in [0,\ 1]$ is a proportionality parameter whereas we indicated with I a general interaction function depending on the pre-interaction states $v,\ v_*$ and on a set of uncertain quantities given by a random variable $z \in \mathbb{R}_+,\ z \sim \Psi$, where $\Psi : \mathbb{R} \to \mathbb{R}_+$ is a probability density function:

$$\mathbb{P}(z \leq \bar{z}) = \int_{-\infty}^{\bar{z}} \Psi(z)dz.$$

It is worth mentioning that (1) is structurally anisotropic. Indeed, if a car behind another one modifies its speed this action does not induce that leading car to go faster or slower. Those assumptions are in agreement with follow-the-leader microscopic models, see [4] for the original microscopic modelling set-up and [21] for further investigation on their kinetic counterpart.

The interaction describes the tendency to update the vehicle speed v taking into account the speed of the other vehicle v_* and an uncertain traffic composition. The term $D(v)\eta$ takes into account stochastic fluctuations due to possible deviations from the deterministic behaviour modelled by the interaction function I. In particular, $\eta \in \mathbb{R}$ is a centred random variable with finite non-zero variance, i.e.

$$\langle \eta \rangle = 0, \qquad \langle \eta^2 \rangle = \sigma^2 > 0,$$

where $\langle \cdot \rangle$ denotes the expectation with respect to the law of the random variable η and σ is the standard deviation of η. The function $D : [0,\ 1] \to \mathbb{R}$ expresses the local relevance of the stochastic fluctuations.

2.1 The Interaction Function

Recently, several interaction rules have been proposed in the literature on kinetic models for traffic flows in the absence of uncertainties. Generally, the interaction is modelled by considering separately the cases of acceleration, i.e. $v \leq W$, or deceleration, i.e. $v > W$, where drivers decide their behaviour depending on a certain quantity W. If W depends on the speed of the other vehicle, i.e. $W = g(v_*)$, then real binary interactions take place, g being a given function acting as a threshold. See, for instance, [6, 9, 13, 27, 30] and also [7] for a review on possible interactions.

Here, taking inspiration from [29], we will consider instead the following interaction function:

$$I(v, v_*; z) = P(\rho; z)(1 - v) + (1 - P(\rho; z))(P(\rho; z)v_* - v), \tag{2}$$

where $P(\rho; z) \in [0, 1]$ is the probability of acceleration. The interaction (2) is a convex combination between the tendency to travel with maximal speed, which is unitary in a dimensionless setting, and the necessity to adapt the speed to a fraction of the speed of the leading vehicle. Notice that (2) synthesises the post-interaction speed as a negotiation between acceleration and deceleration, without any threshold W triggering sharply either of them.

The function $P(\rho; z)$ depends on the dimensionless traffic density $\rho \in [0, 1]$ and on the uncertain quantity z. The form that we will consider in this chapter is the following:

$$P(\rho; z) = (1 - \rho)^z, \qquad z > 0. \tag{3}$$

In particular, traffic flows with heterogeneous classes of vehicles are associated with different exponents of the function $P(\rho; z)$.

Remark 1 A compatibility condition for the interaction rule (1) with the interaction function (2) is that the post-interaction speeds remain in the interval [0, 1]. This can be guaranteed by imposing the following sufficient condition on the fluctuation $D(v)\eta$:

$$|\eta| \le c(1 - \gamma), \qquad cD(v) \le \min\{v, 1 - v\},$$

where $c > 0$ is an arbitrary constant, see [28, 29].

2.2 *Kinetic Description and Equilibria*

Let $f = f(t, v; z)$ be the distribution function of the vehicles travelling with speed $v \in [0, 1]$ at time $t \ge 0$ and belonging to the vehicle class $z \sim \Phi(z)$. Since microscopic interactions are binary and Remark 1 ensures that the post-interaction speeds remain always in [0, 1], we may rely on a Boltzmann-type equation for Maxwellian-like particles for the evolution of f, which is weak form is written as

$$\frac{d}{dt}\int_0^1 \varphi(v)f(t, v; z)\,dv = \frac{1}{2}\int_0^1\int_0^1 \langle\varphi(v') - \varphi(v)\rangle f(t, v; z)f(t, v_*; z)\,dv\,dv_*, \tag{4}$$

where $\varphi : [0, 1] \to \mathbb{R}$ is a test function. We refer the reader to [20] for a detailed derivation of such a kinetic equation for collective phenomena.

From (4) we may obtain information on the evolution of observable quantities, such as the mass of vehicles and their mean speed. In particular, letting $\varphi(v) = 1$ we observe that the mass of the system is conserved since

$$\frac{d}{dt}\int_0^1 f(t, v; z)dv = 0$$

for all $z \in \mathbb{R}_+$. Therefore if at time $t = 0$ the distribution f is a probability density it remains so for all times $t > 0$. Furthermore, for $\varphi(v) = v$ we have

$$\begin{aligned}\frac{d}{dt}V(t; z) &= \frac{\gamma}{2}\int_0^1\int_0^1 I(v, v_*; z)f(t, v; z)f(t, v_*; z)\,dv\,dv_* \\ &= \frac{\gamma}{2}\Big[P(\rho; z)(1-V) - (1-P(\rho; z))^2 V\Big],\end{aligned}$$

where $V(t; z)$ is the uncertain mean speed of the flow. For large times ($t \to +\infty$), we obtain its asymptotic profile which depends now uniquely on the system uncertainty and on the known traffic density ρ:

$$V_\infty(\rho; z) = \frac{P(\rho; z)}{P(\rho; z) + (1-P(\rho; z))^2}. \tag{5}$$

Similarly, for the evolution of the energy we consider $\varphi(v) = v^2$ to obtain

$$\begin{aligned}\frac{dE}{dt} &= \frac{\gamma}{2}\int_0^1\int_0^1 \Big(\gamma I^2(v, v_*; z) + 2vI(v, v_*; z)\Big) f(t, v; z)f(t, v_*; z)\,dv\,dv_* \\ &\quad + \frac{\sigma^2}{2}\int_0^1 D^2(v)f(t, v; z)\,dv \\ &= \frac{\gamma^2P^2}{2}(1+E-2V) + \frac{\gamma^2(1-P)^2}{2}\Big(\Big(P^2+1\Big)E - 2PV^2\Big) \\ &\quad + \gamma P(V-E) + \gamma(1-P)(PV^2-E) \\ &\quad + \frac{\sigma^2}{2}\int_0^1 D^2(v)f(t, v; z)\,dv.\end{aligned}$$

In the zero-diffusion limit $\sigma^2 \to 0^+$ and for a traffic regime $\rho \in (0, 1)$, the energy evolution reduces to

$$\begin{aligned}\frac{dE}{dt} &= \frac{\gamma^2}{2}\Big[P^2(1+E-2V) + (1-P)^2\Big(\Big(P^2+1\Big)E - 2PV^2\Big)\Big] \\ &\quad + \gamma P(V-E) + \gamma(1-P)\Big(PV^2 - E\Big).\end{aligned}$$

For large times, using (5) we have

$$E_\infty(\rho;\, z) = \left(\frac{P(\rho;\, z)}{P(\rho;\, z) + (1 - P(\rho;\, z))^2}\right)^2 = V_\infty^2(\rho;\, z).$$

Therefore, for all $\rho \in (0,\, 1)$ the large time distribution is a Dirac delta $\delta(v - V_\infty(\rho;\, z))$ centred in the z-dependent asymptotic mean speed $V_\infty(\rho;\, z)$. It is interesting to observe that in the traffic regime $\rho \to 0^+$, leading to $P \to 1$ (cf. (3)), the evolution of the energy reduces to

$$\frac{d}{dt}E(t;\, z) = \gamma\left(1 - \frac{\gamma}{2}\right)(1 - E(t;\, z)),$$

and therefore

$$E(t;\, z) = (E(0;\, z) - 1)e^{-\gamma(1-\gamma/2)t} + 1.$$

If $0 < \gamma < 2$, then $E(t;\, z) \to 1$ for every $z \in \mathbb{R}_+$. Since for $\rho \to 0^+$ we also have $V(t;\, z) \to 1^-$, the asymptotic distribution is again a Dirac delta $\delta(v - 1)$, however, independent of z. An analogous remark holds for $\rho \to 1^-$, for which $V(t;\, z) \to 0^+$ and the evolution of the energy is given by

$$\frac{d}{dt}E(t;\, z) = -\gamma\left(1 - \frac{\gamma}{2}\right)E(t;\, z),$$

leading now asymptotically to the Dirac delta $\delta(v)$ again independent of z.

A more detailed analysis of the aggregate behaviour of the kinetic model may be obtained looking at the equilibrium distribution for non-vanishing diffusion. Unfortunately, clear analytical insights are not easy to obtain in general, due to the complexity of the collision operator at the right-hand side of (4). In order to overcome this difficulty, we may, however, rely on the powerful asymptotic method of the quasi-invariant limit, see [2, 25]. The idea is to consider the regime in which the parameters γ, σ^2 of the microscopic interactions are small, so that each interaction produces a small change of speed of the vehicles. At the same time, in order to balance the weakness of the interactions and to observe a trend in the limit $\gamma,\, \sigma^2 \to 0^+$, one increases their rate by introducing the new time scale $\tau := \gamma t/2$ and the scaled kinetic distribution function

$$g(\tau,\, v;\, z) := f\left(\frac{2\tau}{\gamma},\, v;\, z\right),$$

which from (4) is easily seen to solve the following Boltzmann-type equation:

$$\frac{d}{d\tau}\int_0^1 \varphi(v)g(\tau,\, v;\, z)\,dv = \frac{1}{\gamma}\int_0^1\int_0^1 \langle\varphi(v') - \varphi(v)\rangle g(\tau,\, v;\, z)g(\tau,\, v_*;\, z)\,dv\,dv_*.$$

It is possible to prove, see [29], that if η has moments bounded up to the third order, then, in the limit $\gamma, \sigma^2 \to 0^+$ with $\sigma^2/\gamma \to \lambda > 0$, g satisfies the following Fokker-Planck equation with non-constant coefficients:

$$\partial_\tau g = \frac{\lambda}{2}\partial_v^2 \left(D^2(v)g\right) - \partial_v \left((P(1+(1-P)U(\tau;\, z)) - v)g\right), \tag{6}$$

where

$$U(\tau;\, z) := \int_0^1 vg(\tau,\, v;\, z)\,dv = V\left(\frac{2}{\gamma}\tau;\, z\right)$$

denotes the mean speed in the new time scale. Notice that $U(\tau;\, z) \to V_\infty(\rho;\, z)$ for $\tau \to +\infty$, because from the performed time scaling we infer that τ/t is constant for every $\gamma > 0$.

We can now investigate the large time trends of Eq. (6) more easily. Following [29], it can be shown that at the steady state Eq. (6) is such that

$$\frac{\lambda}{2}\partial_v(D^2(v)g_\infty(v;\, z)) - (V_\infty(\rho;\, z) - v)g_\infty(v;\, z) = 0,$$

where we denoted by g_∞ the stationary speed distribution, whose general solution reads

$$g_\infty(v;\, z) = C_{\lambda,\rho,z} \exp\left\{-\frac{2}{\lambda}\int \frac{V_\infty(\rho;\, z) - v}{D^2(v)}dv\right\}, \tag{7}$$

$C_{\lambda,\rho,z} > 0$ being a normalisation constant such that $\int_0^1 g_\infty(v;\, z)dv = 1$ for all z. Depending on the choice of the function $D(v)$ different particular distributions may be obtained, a broad range of which has been investigated in [25].

The empirical speed distributions of traffic are typically supported in the bounded interval $[0,\, 1]$, therefore classical probability densities, such as the normal and the log-normal ones, are not good approximations of the observable stationary profiles. It is worth remarking that the first attempts to fit speed profiles date back to the half of the past century, see [1]. In those original approaches, a deviation of the real data from the standard normal distribution was noticed, in particular, when the traffic density is close to the road capacity, for in that case the speed distribution becomes heavily skewed. More recently, beta distributions have been identified to fit quite well the experimental data of traffic speeds, see [16, 19] for a detailed account of statistical tests validating this conclusion. Interestingly, beta distributions may be obtained from (7) with the choice

$$D(v) = \sqrt{v(1-v)},$$

which produces

$$g_\infty(v;\ z) = C_{\lambda,\rho,z} v^{\frac{2V_\infty(\rho;\,z)}{\lambda}-1}(1-v)^{\frac{2(1-V_\infty(\rho;\,z))}{\lambda}-1}, \tag{8}$$

with

$$C_{\lambda,\rho,z} := \frac{1}{\mathrm{B}\left(\frac{2V_\infty(\rho;\,z)}{\lambda},\ \frac{2(1-V_\infty(\rho;\,z))}{\lambda}\right)}, \tag{9}$$

where $\mathrm{B}(\cdot,\ \cdot)$ is the beta function. Taking advantage of the known formulas for beta-distributed random variables, we easily see that, consistently with the kinetic model, the distribution (8) has mean $V_\infty(\rho;\ z)$ and energy given by

$$\int_0^1 v^2 g_\infty(v;\ z)dv = \frac{V_\infty(\rho;\ z)}{2+\lambda}\left(2V_\infty(\rho;\ z)+\lambda\right).$$

2.3 *Quantities of Interest*

In order to validate our theoretical results by means of experimental data we need to define some quantities of interest to be observed. The advantage of our kinetic approach consists in an analytically closed and sufficiently rich description of the speed profiles emerging at equilibrium, which can be fruitfully compared with the information contained in the measured dataset.

Since the emerging equilibria are affected by the uncertainty brought by the parameter z, it is of paramount importance to define what we may observe if we compare theoretical profiles with experimental data. In view of the mixed traffic conditions, where different vehicles interact and modify their speeds, it is natural to measure expected quantities with respect to the z-uncertainty. Therefore, the reconstructed speed distribution has to be compared with the following expected distribution:

$$\begin{aligned} \bar{g}_\infty(v) &:= \mathbb{E}_z\left[C_{\lambda,\rho,z} v^{\frac{2V_\infty(\rho;\,z)}{\lambda}-1}(1-v)^{\frac{2(1-V_\infty(\rho;\,z))}{\lambda}-1}\right] \\ &= \int_{\mathbb{R}_+} C_{\lambda,\rho,z} v^{\frac{2V_\infty(\rho;\,z)}{\lambda}-1}(1-v)^{\frac{2(1-V_\infty(\rho;\,z))}{\lambda}-1}\Psi(z)\,dz, \end{aligned} \tag{10}$$

where the normalisation constant $C_{\lambda,\rho,z}$ has been defined in (9). This poses the necessity to determine the more suited distribution $\Psi(z)$ that classifies the reaction strengths of the real flow.

Among the most studied diagrams for traffic dynamics, the fundamental diagram summarises macroscopic trends in terms of predicted flow in connection with the recorded density. The fundamental diagram may be obtained from the introduced kinetic modelling by looking at the equilibrium relationship between the traffic density and the z-averaged macroscopic flux of the vehicles, i.e. the mapping $\rho \mapsto \rho \bar{V}_\infty(\rho)$. Then the observable macroscopic trends are given by the following expected quantities:

$$\bar{V}_\infty(\rho) := \mathbb{E}_z\left(V_\infty(\rho;\, z)\right) = \int_{\mathbb{R}_+} V_\infty(\rho;\, z)\Psi(z)\,dz,$$
$$\bar{E}_\infty(\rho) := \mathbb{E}_z(E_\infty(\rho;\, z)) = \int_{\mathbb{R}_+} E_\infty(\rho;\, z)\Psi(z)\,dz. \tag{11}$$

We may also recover the typical scattering observed in empirical fundamental diagrams by looking at the set

$$\left\{(\rho,\, q) \in [0,\, 1] \times \mathbb{R}_+ \,:\, q \in \left[\rho\bar{V}_\infty(\rho) - \rho\varsigma(\rho),\, \rho\bar{V}_\infty(\rho) + \rho\varsigma(\rho)\right]\right\},$$

where $\varsigma^2(\rho)$ is the z-variance of $V(\rho;\, z)$. Indeed, the superposition of different microscopic uncertainties due to different values of z is able to explain the observable scattering in this type of diagrams. We refer the interested reader to [28] for deeper insights into this approach and we mention also [9, 22, 24, 30] for alternative approaches.

3 Description of Traffic Data

In this work we consider data published in [14], which have been recently extracted from 15 videos recorded by 5 cameras in a single traffic direction on the German A3 motorway. The road section is composed by three lanes in each direction with a speed limit of 100 km/h. The videos have been recorded in various traffic conditions, between 7:35 am and 8:00 am, for a total of 8305 recorded vehicles. Each camera covers approximately 100 m of road, and they are spaced in such a way that the total recorded road length is 1 km. Therefore, we may consider the collected data as representative of traffic dynamics in various congestion regimes.

The speeds of the vehicles are recovered out of the microscopic positions in consecutive frames. From time-labelled microscopic data, the evolution of macroscopic quantities characterising the flow can be computed, see [8, 11].

In order to recover the distributions of the microscopic speeds associated with a representative value of the density, we proceed as follows. For each single dataset, corresponding to one video recorded by one camera, we fix a sequence of $M+1$ equally spaced discrete times $\{t_k\}_{k=0}^{M}$, such that $t_{k+1} - t_k =: \delta t$, $t_0 = 0$ and $t_M = t_{\max}$, where $t_{\max}$ is the final observation time, in seconds, in the dataset (here, approximately 1500 s for each video). Then, at each discrete time t_k we count the number of vehicles $N(t_k)$ on the road and define the density as $\tilde{\rho}(t_k) := \frac{N(t_k)}{L}$, for $k = 0, \ldots, M$, where L is the length of the section expressed in the unit length of 1 km. Moreover, we collect all the microscopic speeds of the vehicles on the road at the corresponding discrete time t_k. We take $\delta t = 1$ s and apply this procedure separately to each camera, in order to avoid averaging between very different traffic conditions in different sections of the motorway.

All the computed values of the density $\tilde{\rho}$ are normalised with respect to the maximum allowed density on the road, i.e. the stagnation density $\rho_{\max}$. Since this value is not represented well by the data, we prescribe it as a fixed constant, given by the ratio between the number of lanes and the typical vehicle length of 5 m, plus 50% of additional safety distance, so that

$$\rho_{\max} = \frac{3 \text{ lanes}}{7.5 \text{ m}} = 400 \text{ vehicles/km}.$$

This approach allows us to define representative classes for the densities, identifying values in intervals of length $\frac{1}{10}$. No density levels higher than 0.4 have been observed in this dataset. However, as already noticed in [8], this value is higher than the critical value of the density where a capacity drop in the flux is observed. The experimental distributions are obtained by considering all the microscopic speeds corresponding to a density value belonging to a given density level. The microscopic speeds are normalised with respect to the maximum detected speed in the whole dataset.

It is worth mentioning that the vehicles recorded in [14] have been automatically recognised through a 3D tracking system. Those vehicles can be classified in various classes, spanning from personal cars to bus and trucks with different loads. In particular, 29 types were recognised to represent most of the vehicles in the videos. This natural observation is in agreement with what we introduced in Sect. 2 and leads us to consider heterogeneous traffic conditions for each density levels.

4 Calibration and Results

In this section we show the effects of considering microscopic interactions with uncertainty and compare the extrapolated quantities of interest with reconstructions of real data [14]. In particular, we will focus on the comparison between experimental and theoretical speed distributions.

From data we may distinguish four density regimes $\rho \in \{0.1, 0.2, 0.3, 0.4\}$. For each ρ, several approaches are possible when reconstructing distributions from microscopic quantities, here we opt for the *kernel density estimation*. This technique considers a convolution of the empirical measures associated with the data with a smoothing kernel of given bandwidth $s > 0$. Therefore, if $v_1, \ldots, v_N$ are the microscopic speeds associated with the road density ρ, we consider the probability density function

$$\hat{g}(v) = \frac{1}{Ns}\sum_{i=1}^{N} K\left(\frac{v - v_i}{s^2}\right), \qquad s^2 > 0,$$

being $K(x) := \frac{1}{\sqrt{2\pi}}e^{-\frac{x^2}{2}}$. Other possible approaches are the so-called weighted area rule [10] and standard histograms. The kernel density estimation method may be regarded as a suitable mollification of the histograms.

Once the experimental speed distribution $\hat{g}(v)$ has been reconstructed, we need to estimate the proper uncertainty distribution $\Psi = \Psi(z)$ which makes the theoretical $\bar{g}_\infty(v)$, cf. (10), as consistent as possible with $\hat{g}(v)$. Since mixed traffic conditions with different classes of vehicles are recorded, among the possible uncertainty distributions we may consider the case of a discrete random variable $z \in \{z_1, \ldots, z_n\} \subset \mathbb{R}_+$ with law

$$\mathbb{P}(z = z_k) = \alpha_k \in [0, 1], \qquad \sum_{k=1}^{n} \alpha_k = 1, \tag{12}$$

which leads to

$$\Psi(z) := \sum_{k=1}^{n} \alpha_k \delta(z - z_k),$$

where $\delta(z - z_k)$ is the Dirac delta distribution centred in $z = z_k$. In this case, we have that (10) is

$$\bar{g}_\infty(v) = \sum_{k=1}^{n} \alpha_k C_{\lambda,\rho,z_k} v^{\frac{2V_\infty(\rho;\, z_k)}{\lambda}} (1 - v)^{\frac{2(1-V_\infty(\rho;\, z_k))}{\lambda} - 1}, \tag{13}$$

being the mean speed V_∞ defined in (5). Therefore, in general we obtain observable speed distributions depending on $2n$ parameters, specifically $z_1, \ldots, z_n$, $\alpha_1, \ldots, \alpha_{n-1}$ and λ, since $\alpha_n = 1 - \sum_{k=1}^{n-1} \alpha_k$.

In order to compare (13) with the experimental $\hat{g}$ obtained through the kernel density estimation we solve the following constrained optimisation problem:

$$\min_{(z_1, \ldots, z_n, \alpha_1, \ldots, \alpha_{n-1}, \lambda)} \mathcal{J}(\hat{g}, \bar{g}_\infty), \tag{14}$$

Table 1 Values of the estimated set of parameters $(z_1, z_2, \alpha_1, \lambda)$ obtained through the optimisation procedure (14) relative to four density levels

	Parameters				Cost
ρ	z_1	z_2	α_1	λ	$\mathcal{J}(\hat{g}, \bar{g}_\infty)$
0.1	8.365	8.365	0.500	0.1185	0.2533
0.2	6.475	4.140	0.256	0.1185	0.2874
0.3	4.411	2.741	0.528	0.0806	0.4954
0.4	3.186	2.073	0.425	0.0860	0.7603

In the last column we report the value of the cost functional $\mathcal{J}$ (15)

where $\mathcal{J}$ is the cost functional

$$\mathcal{J}(\hat{g}, \bar{g}_\infty) = \left(\int_0^1 |\hat{g}(v) - \bar{g}_\infty(v)|^2 \, dv \right)^{1/2}, \tag{15}$$

namely the L^2 norm of the difference between $\hat{g}$ and $\bar{g}_\infty$. Problem (14) has to be solved under the constraints

$$\begin{aligned} 0 \le \alpha_k \le 1 \qquad & \forall k = 1, \dots, n-1 \\ z_k \ge 0 \qquad & \forall k = 1, \dots, n-1 \\ \lambda \ge 0. & \end{aligned}$$

This optimisation procedure has been performed through the standard `fmincon` algorithm of MATLAB.

In Table 1 we summarise the parameters obtained in the case $n = 2$, where interactions are characterised by two possible strengths corresponding to the simplified case where two classes of vehicles are considered. From (12) we observe that $\alpha_1 + \alpha_2 = 1$. Since the iterative algorithm used for solving the optimisation problem can be sensitive to the initial guesses $z_1^0, \dots, z_N^0, \alpha_1^0, \dots, \alpha_N^0, \lambda^0$, we have performed a further analysis by solving different optimisation problems spanning uniformly distributed values of $z_1^0, z_2^0, \lambda^0 \in [0.1, 3]$ (5 values), $\alpha_1^0 \in [0.1, 0.9]$ (3 values). From this analysis we have obtained 375 sets of parameters. We have selected the optimal set through the minimisation of the L^1 error between the empirical and the expected mean speed and energy, i.e.

$$\text{Err}_1 = \left| \int_0^1 v\hat{g}(v)\, dv - \bar{V}_\infty(\rho) \right|, \qquad \text{Err}_2 = \left| \int_0^1 v^2\hat{g}(v)\, dv - \bar{E}_\infty(\rho) \right|,$$

where $\bar{V}_\infty, \bar{E}_\infty$ have been defined in (11).

In Fig. 1 we plot the reconstructed distributions in the density regimes $\rho \in \{0.1, 0.2, 0.3, 0.4\}$ together with the $\bar{g}_\infty$ obtained with the optimal parameters in the case of uncertainty of the form (12) and $n = 2$. Remarkably, in free traffic regimes, i.e. for $\rho = 0.1, 0.2$, the single peak appearing in $\hat{g}$ is nicely captured by

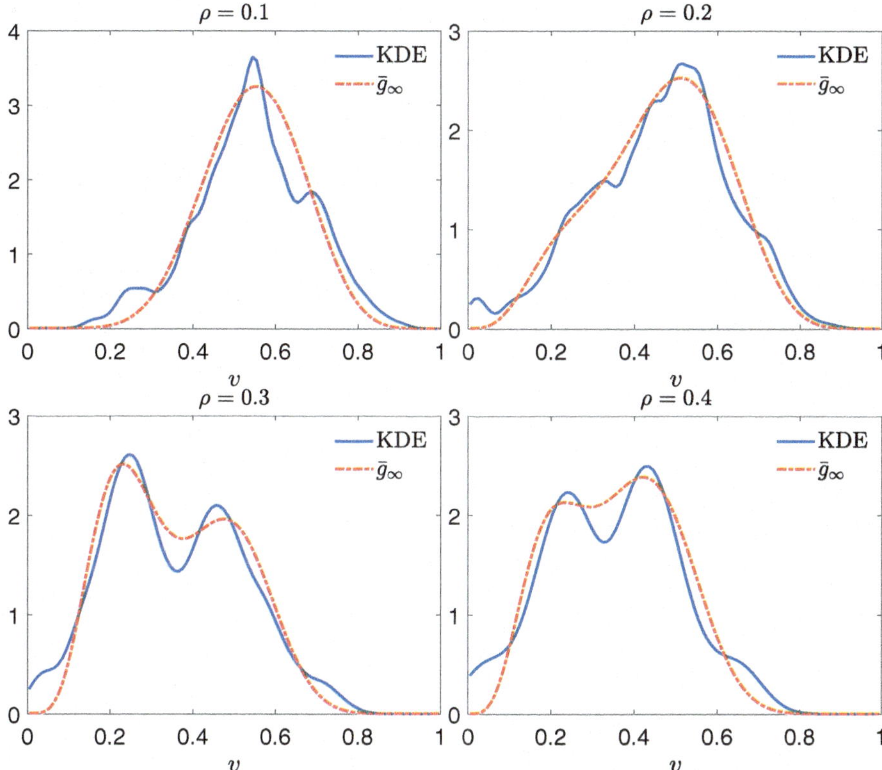

Fig. 1 Comparison between the densities reconstructed from rough data through the kernel density estimation and $\bar{g}_\infty$ defined by the optimal set of parameters reported in Table 1

$\mathbb{E}[g_\infty]$. Furthermore, in congested traffic regimes, i.e. for $\rho \in \{0.3, 0.4\}$, $\hat{g}$ shows a bimodal trend which is in turn nicely captured by $\mathbb{E}[g_\infty]$.

To further validate the proposed approach, in Table 2 we report the values of the first and second moments, namely the mean speed and the energy, extrapolated from the empirical and theoretical distributions $\hat{g}$ and $\bar{g}_\infty$, respectively. In particular, $\bar{V}_\infty$

Table 2 Values of the extrapolated first and second order moments of the distributions $\hat{g}$ and $\bar{g}_\infty$, the latter being the optimised one with the parameters given in Table 1

	Extrapolated moments			
ρ	$\hat{V}$	$\bar{V}_\infty$	$\hat{E}$	$\bar{E}_\infty$
0.1	0.6009	0.6016	0.3498	0.3475
0.2	0.5132	0.5132	0.2673	0.2610
0.3	0.4010	0.3996	0.1746	0.1706
0.4	0.3975	0.3929	0.1691	0.1626

and $\bar{E}_\infty$ defined in (11) are compared with

$$\hat{V} = \int_0^1 v\hat{g}(v)\,dv, \qquad \hat{E} = \int_0^1 v^2\hat{g}(v)\,dv.$$

Interestingly, we observe that the comparison gives perfectly consistent results for all densities.

5 Conclusions

In this work we have proposed an approach for the effective reconstruction of traffic speed distributions starting from the assessment of microscopic rough data. The techniques here developed are rooted in the statistical framework of the Boltzmann-type kinetic theory for multi-agent systems [20]. In particular, the microscopic interactions among the vehicles are assumed to be binary and are defined starting from basic assumptions on the driver behaviour consistent with follow-the-leader-type dynamics. The additional formalism of the uncertainty quantification has allowed us to consider real traffic flows, in which mixed conditions are often observed due e.g. to the simultaneous presence of different types of vehicles. We have translated this feature in uncertain microscopic interactions [26], the uncertain parameter z being one which affects the reaction strength of the vehicles in acceleration/deceleration.

From our kinetic approach, in particular in the asymptotic regime of the quasi-invariant interactions [25], we have been able to compute analytical equilibrium speed distributions, which fit nicely the empirically interpolated beta distributions [16, 19] and maintain also an explicit dependence on the uncertainty parameter z. By estimating the statistical distribution of z via an optimisation procedure grounded on real traffic data recorded on the German A3 motorway [14], we have recovered also more complex speed distributions, such as e.g. bimodal ones emerging in medium density traffic regimes, as the z-averaged superposition of "elementary" beta distributions.

We believe that these results pave the way to a physically sound and mathematically consistent procedure for the reconstruction and quantification of microscopic uncertainties naturally present in collective phenomena, which may have a considerable impact at larger scales.

Acknowledgments This research was partially supported by the Italian Ministry of Education, University and Research (MIUR) through the "Dipartimenti di Eccellenza" Programme (2018–2022)—Department of Mathematical Sciences "G. L. Lagrange", Politecnico di Torino (CUP:E11G18000350001) and Department of Mathematics "F. Casorati", University of Pavia; and through the PRIN 2017 project (No. 2017KKJP4X) "Innovative numerical methods for evolutionary partial differential equations and applications".

This work is also part of the activities of the Starting Grant "Attracting Excellent Professors" funded by "Compagnia di San Paolo" (Torino) and promoted by Politecnico di Torino.

A.T. and M.Z. are members of GNFM (Gruppo Nazionale per la Fisica Matematica) of INdAM (Istituto Nazionale di Alta Matematica), Italy.

The research of M.H. and G.V. is funded by the Deutsche Forschungsgemeinschaft (DFG, German Research Foundation) under Germany's Excellence Strategy—EXC-2023 Internet of Production—390621612.

M.H. and G.V. acknowledge the ISAC institute at RWTH Aachen, Prof. M. Oeser, Dr. A. Fazekas, MSc. M. Berghaus and MSc. E. Kalló for kindly providing the trajectory data within the DFG project "Basic Evaluation for Simulation-Based Crash-Risk-Models: Multi-Scale Modelling Using Dynamic Traffic Flow States".

References

1. D.S. Berry, D.M. Belmont, Distribution of vehicle speeds and travel times, in *Proceedings of the Second Berkeley Symposium on Mathematical Statistics and Probability* (University of California Press, Berkeley, 1951), pp. 589–602
2. S. Cordier, L. Pareschi, G. Toscani, On a kinetic model for a simple market economy. J. Stat. Phys. **120**(1), 253–277 (2005)
3. G. Dimarco, L. Pareschi, M. Zanella, Uncertainty quantification for kinetic models in socio-economic and life sciences, in *Uncertainty Quantification for Hyperbolic and Kinetic Equations*. SEMA-SIMAI Springer Series, vol. 14, ed. by S. Jin, L. Pareschi (Springer, Berlin, 2017), pp. 151–191
4. D.C. Gazis, R. Herman, R.W. Rothery, Nonlinear follow-the-leader models of traffic flow. Oper. Res. **9**(4), 545–567 (1961)
5. *Global Status Report on Road Safety*. Technical report (World Health Organization, Switzerland, 2018)
6. M. Günter, A. Klar, T. Materne, R. Wegener, An explicitly solvable kinetic model for vehicular traffic and associated macroscopic equations. Math. Comp. Model. **35**(5–6), 591–606 (2002)
7. M. Herty, L. Pareschi, Fokker-Planck asymptotics for traffic flow models. Kinet. Relat. Mod. **3**(1), 165–179 (2010)
8. M. Herty, A. Fazekas, G. Visconti, A two-dimensional data-driven model for traffic flow on highways. Netw. Heterog. Media **13**(2), 217–240 (2018)
9. M. Herty, A. Tosin, G. Viconti, M. Zanella, Hybrid stochastic kinetic description of two-dimensional traffic dynamics. SIAM J. Appl. Math. **78**(5), 2737–2762 (2018)
10. R.W. Hockney, J.W. Eastwook, *Computer Simulation using Particles* (McGraw Hill International Book Co., New York, 1981)
11. S.P. Hoogendoorn, Traffic flow theory and simulation, in *Lecture notes CT4821* (Delft University of Technology, New York, 2007)
12. J. Hu, S. Jin, Uncertainty quantification for kinetic equations, in *Uncertainty Quantification for Hyperbolic and Kinetic Equations*. SEMA-SIMAI Springer Series, vol. 14, ed. by S. Jin, L. Pareschi (Springer, Berlin, 2017), pp. 193–229
13. R. Illner, A. Klar, H. Lange, A. Unterreiter, R. Wegener, A kinetic model for vehicular traffic: existence of stationary solutions. J. Math. Anal. Appl. **237**, 622–643 (1999)
14. E. Kallo, A. Fazekas, S. Lamberty, M. Oeser, Microscopic traffic data obtained from videos recorded on a German motorway. Mendeley Data **v1**, 7 (2019)
15. A. Klar, R. Wegener, Enskog-like models for vehicular traffic. J. Stat. Phys. **87**(1–2), 91–114 (1997)
16. A.K. Maurya, S. Das, S. Dey, S. Nama, Study on speed and time-headway distributions on two-lane bidirectional road in heterogeneous traffic condition. Transp. Res. Proc. **17**, 428–437 (2016)
17. C.R. Munigety, Modelling behavioural interactions of drivers in mixed traffic conditions. J. Traffic Transp. Eng. **5**(4), 284–295 (2018)

18. C.R. Munigety, T.V. Mathew, Towards behavioral modeling of drivers in mixed traffic conditions. Transp. Dev. Econom. **2**(6) (2016)
19. D. Ni, H.K. Hsieh, T. Jiang, Modeling phase diagrams as stochastic processes with application in vehicular traffic flow. Appl. Math. Model. **53**, 106–117 (2018)
20. L. Pareschi, G. Toscani, *Interacting Multiagent Systems: Kinetic Equations and Monte Carlo Methods* (Oxford University, Oxford, 2013)
21. B. Piccoli, A. Tosin, M. Zanella, Model-based assessment of the impact of driver-assist vehicles using kinetic theory. Preprint arXiv:1911.04911
22. G. Puppo, M. Semplice, A. Tosin, G. Visconti, Fundamental diagrams in traffic flow: the case of heterogeneous kinetic models. Commun. Math. Sci. **14**(3), 643–669 (2016)
23. G. Puppo, M. Semplice, A. Tosin, G. Visconti, Analysis of a multi-population kinetic model for traffic flow. Commun. Math. Sci. **15**(2), 379–412 (2017)
24. B. Seibold, M.R. Flynn, A.R. Kasimov, R.R. Rosales, Constructing set-valued fundamental diagrams from Jamiton solutions in second order traffic models. Netw. Heterog. Media **8**(3), 745–772 (2013)
25. G. Toscani, Kinetic models of opinion formation. Commun. Math. Sci. **4**(3), 481–496 (2006)
26. A. Tosin, M. Zanella, Boltzmann-type models with uncertain binary interactions. Commun. Math. Sci. **16**(4), 963–985 (2018)
27. A. Tosin, M. Zanella, Control strategies for road risk mitigation in kinetic traffic modelling. IFAC-PapersOnLine **51**(9), 67–72 (2018)
28. A. Tosin, M. Zanella, Uncertainty damping in kinetic traffic models by driver-assist controls. Math. Control Relat. Fields (2019). Preprint arXiv:1904.00257
29. A. Tosin, M. Zanella, Kinetic-controlled hydrodynamics for traffic models with driver-assist vehicles. Multiscale Model. Simul. **17**(2), 716–749 (2019)
30. G. Visconti, M. Herty, G. Puppo, A. Tosin, Multivalued fundamental diagrams of traffic flow in the kinetic Fokker-Planck limit. Multiscale Model. Simul. **15**, 1267–1293 (2017)
31. D. Xiu, *Numerical Methods for Stochastic Computations*. (Princeton University, Princeton, 2010)

From Kinetic to Macroscopic Models and Back

Michael Herty, Gabriella Puppo, and Giuseppe Visconti

Abstract We study kinetic models for traffic flow characterized by the property of producing backward propagating waves. These waves may be identified with the phenomenon of stop-and-go waves typically observed on highways. In particular, a refined modeling of the space of the microscopic speeds and of the interaction rate in the kinetic model allows to obtain weakly unstable backward propagating waves in dense traffic, without relying on non-local terms or multi–valued fundamental diagrams. A stability analysis of these waves is carried out using the Chapman-Enskog expansion. This leads to a BGK-type model derived as the mesoscopic limit of a Follow-The-Leader or Bando model, and its macroscopic limit belongs to the class of second-order Aw-Rascle and Zhang models.

1 Introduction

There are mainly three modeling scales in the mathematical description of vehicular traffic flow. The microscopic scale is based on the prediction of trajectories of individual vehicles by systems of ordinary differential equations. The macroscopic scale is based on the assumption that traffic flow behaves like a fluid where individual vehicles cannot be identified, but a macroscopic conservation law for the number of vehicles rules the dynamics. Here, the flow is represented by a density function and evolves in space and time by transport equations. The intermediate scale is the mesoscopic scale. Here, kinetic equations govern the dynamics. Those equations are characterized by a statistical description of the microscopic states of vehicles but, at the same time, still provide the macroscopic aggregate representation

M. Herty · G. Visconti
RWTH Aachen University, Aachen, Germany
e-mail: herty@igpm.rwth-aachen.de; visconti@igpm.rwth-aachen.de

G. Puppo (✉)
"La Sapienza" Università di Roma, Roma, Italy
e-mail: gabriella.puppo@uniroma1.it

G. Puppo, A. Tosin (eds.), *Mathematical Descriptions of Traffic Flow: Micro, Macro and Kinetic Models*, SEMA SIMAI Springer Series 12,
https://doi.org/10.1007/978-3-030-66560-9_2

of traffic flow, linking collective dynamics to interactions among vehicles at a smaller microscopic scale.

In the present chapter we study non-homogeneous kinetic models for vehicular traffic flow. In particular, we investigate the common and well-established idea that non-local terms are necessary to observe backward propagation of waves in dense traffic [13]. We show that the model in [17] naturally encloses backward propagating waves, although these waves may be unstable. We include a first stabilization term including the effect of uncertainty in the braking rate [19]. We propose a more refined choice of the interaction rate which allows us to obtain weakly unstable waves propagating back in congested traffic situations without considering non-local terms. More precisely, drawing inspiration from the Knudsen number in kinetic gas-dynamics, we prescribe the interaction rate as a suitable function of the density and its space derivative. The backward propagating waves may still be unstable in the sense that they may exhibit unbounded growth in time. We study the appearance of these instabilities by considering BGK-type (Bhatnagar, Gross and Krook [4]) models in the limit of constant but sufficiently small interaction rates. In this regime it has been shown in [5] that Enskog-like terms provide a stabilization effect. However, in that work the stabilization is unfortunately too strong and it implies that, for example, stop-and-go waves will **not** occur. Following the approach introduced in [11], we derive a weakly unstable BGK model modifying the design of the space of microscopic speeds. Further, we obtain by suitable limits from this mesoscopic representation a microscopic follow-the-leader [9] or Bando [3] model, and a macroscopic Aw-Rascle [2] and Zhang [21] type model.

The results of this work show that multivalued fundamental diagrams and non-local effects are already naturally included in a kinetic traffic model, provided the relaxation rate depends on the space derivative of the density, and one considers non-equilibrium effects. Therefore, there is no need to add non-local effects and multivalued desired velocities to a kinetic model in order to explain observed phenomena in traffic flow. These features in fact are already present in standard kinetic models as non-equilibrium effects.

The chapter is organized as follows. In Sect. 2 we introduce Boltzmann-like kinetic models for traffic flow characterized by binary interactions with over-braking, and we provide an experimental evidence of the backward propagation of waves in dense traffic. In Sect. 3 we analyze the stability of these waves by a Chapman-Enskog expansion of the BGK approximation of the full kinetic model, and we compare the results with the Chapman-Enskog expansion of the BGK model in [5] and of the Aw-Rascle and Zhang model. Finally, in Sect. 4 we derive a modified version of the BGK model, as in [11], and analyze the stability in the case of interactions with over-braking. In Sect. 5 we discuss results and future perspectives.

2 Backward Propagation of Waves in a Kinetic Traffic Model

A kinetic traffic model for the mesoscopic scale reads as follows

$$\partial_t f(x, v, t) + v\partial_x f(x, v, t) = \frac{1}{\varepsilon} Q[f, f](x, v, t), \tag{1}$$

where $f(x, v, t) : \mathbb{R} \times [0, V_M] \times \mathbb{R}^+ \to \mathbb{R}^+$ is the mass distribution function of the flow and the local traffic density $\rho(t, x)$ is given by

$$\rho(x, t) = \int_0^{V_M} f(x, v, t)\mathrm{d}v. \tag{2}$$

We suppose that the space of possible microscopic speeds of the vehicles is bounded by zero and a maximum speed V_M. Further, we assume that $f(x, v, t = 0)$ is such that density is limited by a maximum density $\rho_M = \int f(x, v, 0)dv < \infty$. Throughout this work, we consider dimensionless quantities and normalize for simplicity $V_M = 1$ and $\rho_M = 1$. The source term in (1) is commonly called collision kernel, in analogy to kinetic models for gas-dynamics, and it models the change of f due to the microscopic interactions among vehicles. $Q[f, f]$ can be modeled as a non-linear integral operator, typical of Boltzmann-type kernels, or as a linear operator, typical of BGK-type kernels. The quantity ε is positive, and yields a relaxation rate weighting the relative strength between the convective term and the source term. It is related to the Knudsen number in fluid dynamics. Generally, ε can be a function of density ρ, and possibly of its spatial derivative. Here, we consider both the case $\varepsilon = \varepsilon(\rho, \partial_x\rho)$ and the case of a constant rate ε.

2.1 A Boltzmann-Type Kinetic Model for Traffic Flow

In the collision operator we model the adaptation of vehicles' speeds by binary car-to-car interaction. This behavior is typical for real-world traffic where usually a driver reacts to the actions of the vehicle in front. To describe the interactions we split the operator $Q[f, f]$ in the difference between a gain term and a loss term. The former accounts for the increase of $f(x, v, t)$ when a vehicle with velocity v_* interacts with a leading vehicle with speed v^*, emerging with speed v as a result of the interaction. The latter accounts for the decrease of $f(x, v, t)$ if a vehicle with velocity v interacts with a vehicle with speed v^*, emerging with speed different from v as a result of the interaction. We assume that the velocity of the leading vehicle remains always unchanged. More specifically,

$$Q[f,f](x,v,t)=\int_0^1\int_0^1 \mathcal{P}(v_*\to v|v^*;\rho)f(x,v_*,t)f(x,v^*,t)\mathrm{d}v_*\mathrm{d}v^* \\ -f(x,v,t)\int_0^1 f(x,v^*,t)\mathrm{d}v^*. \tag{3}$$

The core of a kinetic model is the definition of the operator $\mathcal{P}(v_*\to v|v^*;\rho)$ that prescribes, in a probabilistic way, the resulting speed of a vehicle after interacting with a leading vehicle. The kinetic model for traffic flow studied here is based on the following interaction rules:

$$\mathcal{P}(v_*\to v|v^*;\rho)=\begin{cases} P(\rho)\,\delta_{\min\{v_*+\Delta_a,V_M\}}(v)+(1-P(\rho))\,\delta_{\max\{v_*-\Delta_b,0\}}(v) & v_*\le v^* \\ P(\rho)\,\delta_{\min\{v_*+\Delta_a,V_M\}}(v)+(1-P(\rho))\,\delta_{\max\{v^*-\Delta_b,0\}}(v) & v_*>v^*, \end{cases} \tag{4}$$

where $P(\rho)\in[0,1]$ is a decreasing function of the density modeling the probability of accelerating. The parameters Δ_a and Δ_b are the acceleration and the braking parameters, respectively, where Δ_a is the instantaneous physical acceleration of a vehicle. The parameter Δ_b instead corresponds to an uncertainty in the estimate of the other vehicle's speed. Indeed, $\Delta_b=0$ corresponds to no uncertainty: the vehicle has an exact perception of velocities, and therefore is able to maintain its own speed $v=v_*$ when it interacts with a faster vehicle (i.e., when $v_*<v^*$), while it can brake exactly to the speed $v=v^*$ in case a slower vehicle is ahead (i.e., when $v_*>v^*$). For $\Delta_b=0$ the model [17] is recovered. More details on the case $\Delta_b>0$ can be found in [19]. Note that the model is continuous across the line $v_*=v^*$, ensuring well-posedness, see [16], and that mass conservation holds:

$$\mathcal{P}(v_*\to v|v^*;\rho)\ge 0,\qquad \int_0^1 \mathcal{P}(v_*\to v|v^*;\rho)\mathrm{d}v=1.$$

In the space homogeneous case $f=f(v,t)$, the model (1) reduces to a relaxation to equilibrium which is characterized by a function $M_f(v;\rho)$ such that $Q[M_f,M_f]=0$. In analogy to kinetic models for rarefied gas-dynamics, the function M_f will be called Maxwellian and it allows us to define the flux and the mean speed of vehicles at equilibrium as

$$F_{\mathrm{eq}}(\rho)=\big(\rho U_{\mathrm{eq}}(\rho)\big)=\int_0^1 vM_f(v;\rho)\mathrm{d}v,\quad U_{\mathrm{eq}}(\rho)=\frac{1}{\rho}\int_0^1 vM_f(v;\rho)\mathrm{d}v. \tag{5}$$

For $\Delta_b=0$ it is proven, cf. [17], that stable equilibria are uniquely defined by the local density. Moreover, the Maxwellian is a known function of v, it can be explicitly computed, and depends on x and t only through the local density $\rho(x,t)$.

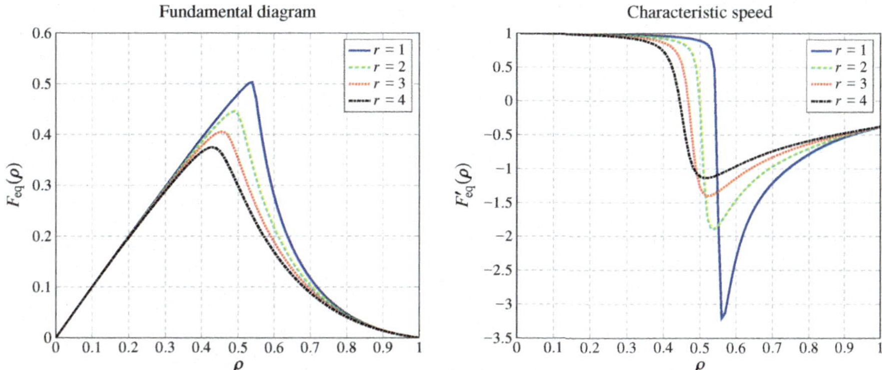

Fig. 1 Fundamental diagrams (left) and characteristic speed (right) with 48 discrete microscopic speeds, $\Delta_a = \frac{1}{4}$ and $\Delta_b = \frac{\Delta_a}{r}$, $r = 1, 2, 3, 4$

Further, in the space homogeneous case, the density is a scalar parameter fixed at the initial time. However, unstable equilibria may also occur, for which the Maxwellian depends not only on ρ but also on the initial distribution $f(x, v, t = 0)$. These equilibria are unstable under perturbation of the initial datum. The Maxwellian corresponding to the stable equilibria is a finite weighted sum of Dirac's functions for any initial distribution. If the braking uncertainty $\Delta_b \neq 0$, it has been shown in [19] that the equilibria corresponding to a given density are unique, and all equilibria are stable.

In Fig. 1 we show the equilibrium flux $F_{\text{eq}}(\rho)$, also known as fundamental diagram, and the characteristic speed $F'_{\text{eq}}(\rho)$ obtained numerically by using 48 discrete equidistant discretization points in the velocity phase space, a fixed value of the acceleration parameter $\Delta_a = \frac{V_M}{4} = \frac{1}{4}$ and different values of the uncertainty Δ_b such that $r = \frac{\Delta_a}{\Delta_b} = 1, 2, 3, 4$. In all cases, the fundamental diagram is characterized by two phases. For low values of ρ the flux is linear in ρ. This is the phase of free flow. For larger values of ρ, the role of the interactions increases, and the flux decreases. This corresponds to the congested phase of traffic flow. The value of the density for which the change between congested and free flow occurs is called *critical density*. Note that the road capacity, i.e., the maximum of the flux, decreases as the uncertainty Δ_b increases.

2.2 *Propagation of Waves*

Integrating equation (1) in velocity space, the right-hand side vanishes because of mass conservation, and one obtains the evolution equation for the density

$$\partial_t \rho(x, t) + \partial_x F(x, t; f) = 0, \quad F(x, t; f) = \int_0^1 v f(x, v, t) \mathrm{d}v, \tag{6}$$

where F is the macroscopic flux obtained through the kinetic model. If the system approaches equilibrium, $f \rightarrow M_f$, and the macroscopic equation reduces to the equilibrium equation

$$\partial_t \rho(x,t) + \partial_x F_{\text{eq}}(\rho(x,t)) = 0, \quad F_{\text{eq}}(\rho(x,t)) = \int_0^1 v M_f(v;\rho) \mathrm{d}v. \tag{7}$$

Since the Maxwellian is defined by ρ, the equilibrium equation (7) is closed, and it is a well-defined scalar conservation law where the flux function $F_{\text{eq}}(\rho)$ is the fundamental diagram. On the other hand, when the system is not at equilibrium, the macroscopic equation (6) is still coupled to the kinetic equation (1).

At the mesoscopic scale, the relaxation speed defined by ε plays a crucial role since, balancing the weight between the convection and the source term, it allows us to define the regimes of the kinetic model. If we allow for $\varepsilon = 0$, i.e., we suppose that the interactions are so frequent to instantaneously relax f to the local equilibrium distribution M_f, we are in the so-called equilibrium flow regime where (1) reduces to the conservation law for the density (7). Instead, we expect that if ε is small, but not vanishing, then we are either in a regime where the kinetic equation (1) reduces to a *perturbed* continuity equation (7) or where the kinetic equation can be approximated by an extended continuum hydrodynamic system of equations as, for example, the Aw-Rascle and Zhang model. For $\varepsilon \asymp 1$, but not too large, we are in the kinetic regime and finally for $\varepsilon \gg 1$ we obtain the regime of the collision-less kinetic equation where the convective term dominates.

In regimes characterized by a small value of ε, we expect that the conservation law (7) should provide a good approximation to the behavior of the solution; in particular smooth waves should travel along the characteristics given by $\partial_\rho F_{\text{eq}}(\rho)$. Thus, looking at the right panel of Fig. 1, we expect that signals move towards the right in the free flow phase and towards the left in the congested flow phase.

However, in the kinetic regime where $\varepsilon >> 1$ signals should always propagate towards the right since the microscopic velocities in traffic are non-negative. This happens also for congested traffic regimes, because the characteristics in the transport term coincide with the microscopic speeds. As observed in [13] this can be seen by computing the implicit solution to (1)

$$f(x,v,t) = f(x - vt, v, t = 0) + \int_0^t Q[f,f](x + v(s-t), v, s) \mathrm{d}s.$$

The distribution function f at x and t depends only on the distribution function at the values $y \le x$ and $s \le t$, since v is non–negative. Thus, apparently, traffic jams in dense flow are not allowed to travel backwards. Several models were introduced in the mathematical literature [8, 13] trying to overcome this drawback.

Numerical evidence suggests strongly that this picture is naive, and that the interaction of the source term, given by the collision operator, and the transport term, is more subtle. We observe instead a smooth transition between the solutions of the equilibrium equation, where signals move backward in congested flow, and solutions of the kinetic equation. Here too in fact the propagation speed of smooth waves can be negative.

To illustrate this point, we show the evolution of the solution of the kinetic model (1) in a few typical cases. In particular, we consider propagating a smooth perturbation in the density

$$\rho_0(x) = a + be^{-8x^2}$$

and periodic boundary conditions. The initial distribution is Maxwellian. The solution is computed with a first-order numerical method, using the local Lax Friedrich's flux. The choice of the numerical flux is crucial: a standard upwind flux, computed following the characteristics of the transport term, would in fact be unstable, in the congested phase, because the direction of the flow *does not* coincide with the direction of the characteristics. Since the collision term becomes stiff for small ε, we penalize the collision term with a BGK operator, as in [7].

We use 4 discrete speeds with $\Delta_a = \Delta_b = \frac{1}{4}$; space is discretized by 200 cells and the final time is $t_f = 1$, while $\varepsilon = 0.01$. The solution is shown at different times, starting from the blue curve at $t = 0$, and ending with the magenta thick profile, at $t = 1$. In the left panel of Fig. 2, we take $a = 0.2$ and $b = 0.2$. The perturbation in the density is below the critical density. Thus, the density profile moves towards the right, as it would occur also in the equilibrium equation. The shape of the initial data is deformed mainly by numerical diffusion, because the flux is almost linear. In the right panel of Fig. 2, we choose $a = 0.7$ and $b = 0.2$, so that the initial perturbation has the same amplitude as before, but it occurs on the dense traffic regime. Now, we observe propagation of the wave towards the left, although the characteristics point towards the right. This means that the propagation speed is governed by the interaction between the collision kernel and the transport term, which reproduces the behavior of the fundamental diagram of the equilibrium equation, where indeed we observe negative characteristics. Note that the height of

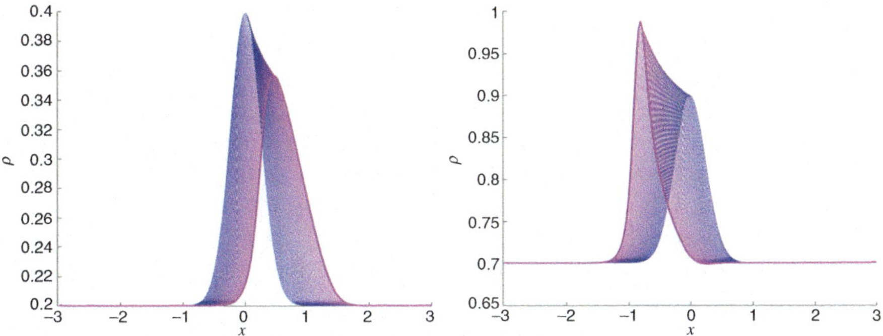

Fig. 2 Time evolution of a density bump in the free flow phase (left) and in the congested flow phase (right). The initial condition is drawn in blue, and the solution shades towards magenta, as time increases

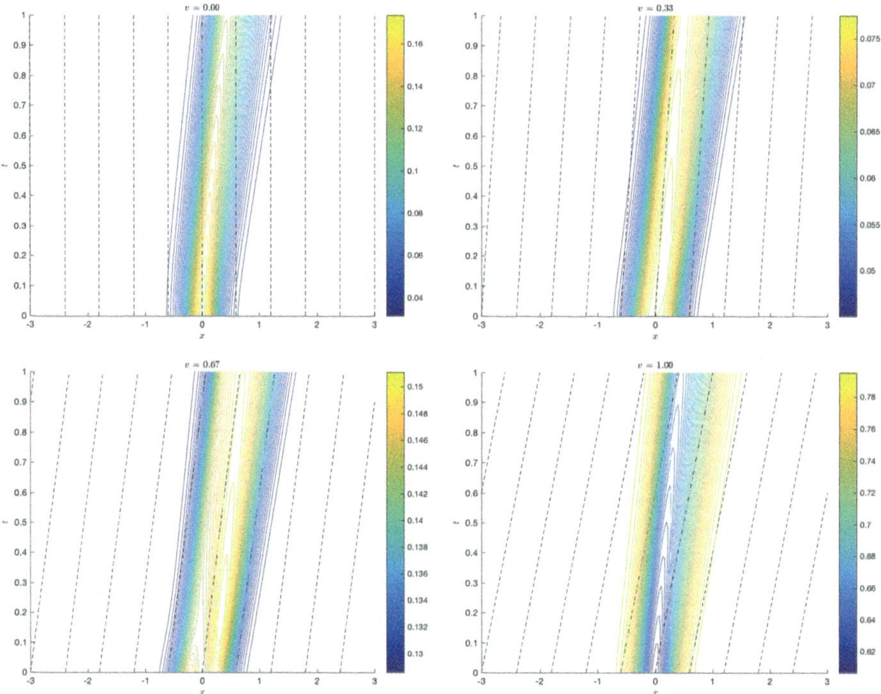

Fig. 3 Space-time evolution of the distribution function for each fixed value of the microscopic speed v, superposed to the corresponding characteristic speed (black dashed lines), during the time evolution of the density bump in free flow

the density peak now *increases* with time: the solution has the correct propagation speed, but it is unstable.

These considerations are further investigated by looking into the distribution function of the kinetic model. We draw contour plots of the space-time behavior of f, for each fixed value of the microscopic speed v. Since we are considering 4 microscopic velocities, we obtain 4 different plots. In the plots, we compare the time development of the solution f with the corresponding characteristic speed of the transport term, drawn with parallel dashed black lines. In the case of the density profile in the free flow phase, we see that the signal propagates towards the right and along characteristics, Fig. 3. Instead, in the case of the density profile in the congested phase, it is clear that the signal propagates towards the left and *across* characteristics, see Fig. 4. Thus the information on the propagation is contained in the interaction of the collision kernel and the transport term, rather than in the convective term alone.

A constant choice of ε, however, is not satisfactory. In fact, in analogy with the Knudsen number in gas models, ε should be a decreasing function of the density. In this way, ε becomes large in the free flow phase since the interactions are less frequent and the convective term rules the dynamics. On the contrary, ε should

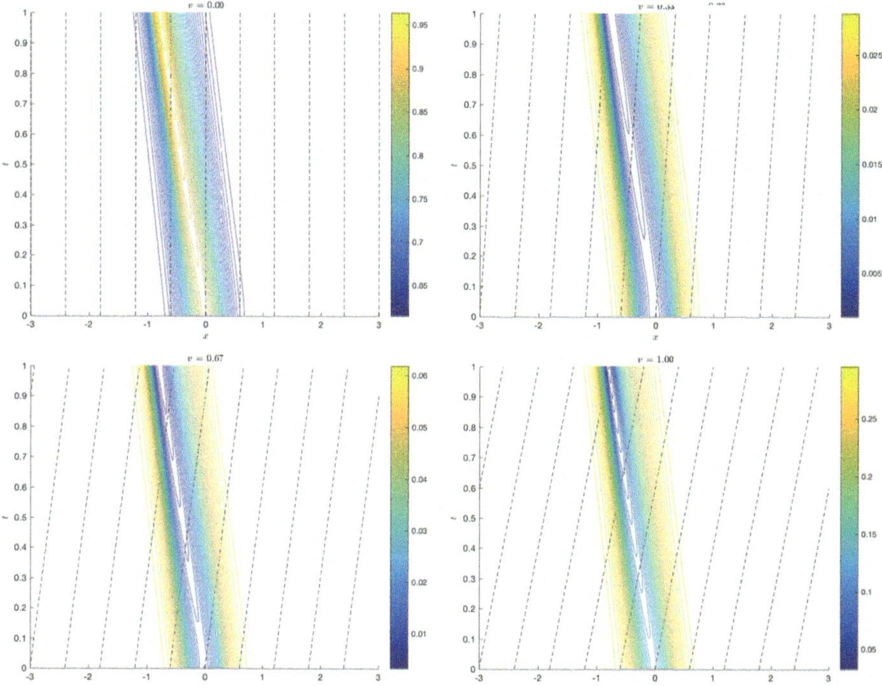

Fig. 4 Space-time evolution of the distribution function for each fixed value of the microscopic speed v, superposed to the corresponding characteristic speed (black dashed lines), during the time evolution of the density bump in congested flow

become small when the density increases, since the relaxation towards equilibrium should be fast when ρ is high and interactions among vehicles are dominant. Further, we also expect that ε should decrease when the traffic thickens, i.e., when ρ_x is large and positive. A choice respecting this argument is

$$\varepsilon(\rho, \rho_x) = \frac{1}{\max\left\{\frac{1}{1-\min\{\rho,\varepsilon_0\}^2}, 1 + \left(\max\{\rho_x, 0\}\right)^2\right\}}, \tag{8}$$

where ε_0 is a threshold to prevent division by zero. The dependence on $\max(\rho_x, 0)$ is crucial to prevent overshoots above the maximum density $\rho_M = 1$, when the density profile is very steep. This might happen if the density increases sharply, as when a fast, low density traffic impinges against a slow congested region. In this case, the presence of $\partial_x \rho$ accounts for the need to look ahead. It replaces the non-locality of the collision term introduced in [13].

A comparison between a fixed ε and the variable collision time of (8) is shown in Fig. 5. The top part of the figure contains the evolution of the high density profile with $a = 0.7$ and $b = 0.2$ up to time $t = 10$. We see that with the variable

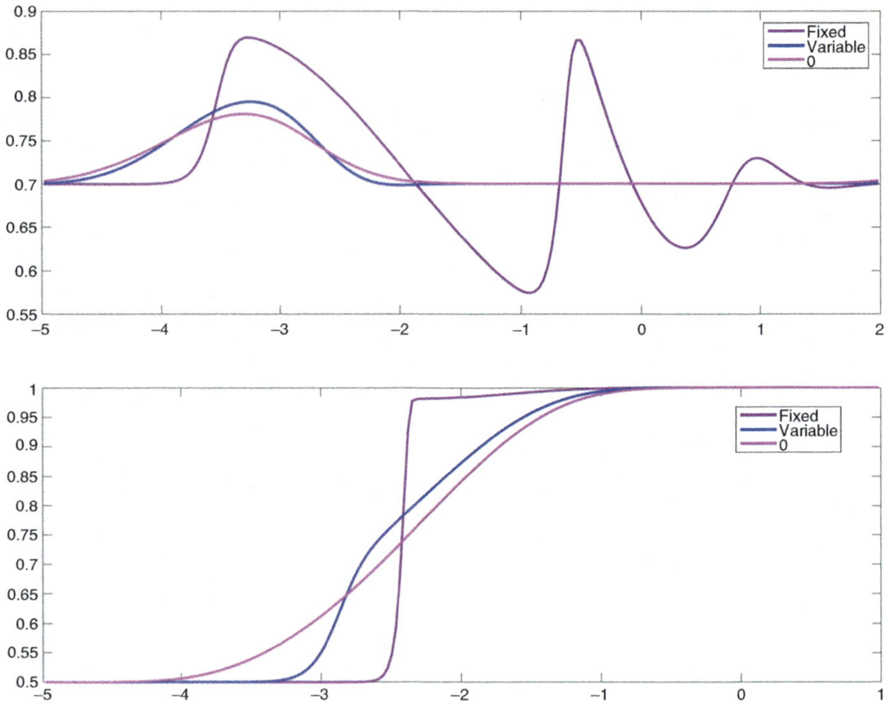

Fig. 5 Comparison of kinetic solutions with variable ε, as in Eq. (8), with fixed $\varepsilon = 0.01$, and with the equilibrium solution $\varepsilon = 0$. Top: solution with a smooth profile in the congested regime. Bottom: solution of a Riemann Problem, corresponding to a red light

collision time, the profile propagates to the left, developing waves which resemble stop-and-go waves. The fixed value of $\varepsilon = 0.01$ prevents the developing of these waves, because the relaxation rate is very strong even when the interaction should be weak. As a comparison, we also show the solution obtained with the equilibrium equation (7).

The bottom part of the figure shows the solution obtained for a Riemann problem mimicking a stream of low density traffic impinging against a queue. Here, the kinetic solution with variable ε develops correctly a shock wave, while the equilibrium solution yields a smooth wave, because, in the congested regime, the fundamental diagram of (7) is convex.

3 Analysis of Instabilities via Chapman-Enskog Expansion

The presence of instabilities is investigated using a formal Chapman-Enskog expansion. For the sake of simplicity, the analysis is performed using the BGK approximation of the Boltzmann-type collision kernel (3), for constant but small values of ε. Unstable waves are present also in this linearized setting [11].

3.1 BGK Approximations with and Without Non-local Terms

The BGK approximation to the kinetic model (1) reads

$$\partial_t f(x, v, t) + v\partial_x f(x, v, t) = \frac{1}{\varepsilon}\left(M_f(v; \rho) - f(x, v, t)\right). \tag{9}$$

The BGK model is an approximation of the full kinetic equation, which holds for small values of ε. In fact, (1) and (9) have, by construction, the same equilibrium solution. This further motivates the use of the BGK approximation to investigate the appearance of instabilities in dense traffic, i.e., in the regime of large densities and small ε.

The Chapman-Enskog expansion allows us to study the behavior of (9) when f is a first-order perturbation in ε around the equilibrium distribution $M_f(v; \rho)$. In particular, we consider fixed and small values of ε. Then, plugging the expansion

$$f(x, v, t) = M_f(v; \rho) + \varepsilon f_1(x, v, t), \quad \text{with} \int_0^1 f_1(x, v, t)\mathrm{d}v = 0,$$

into (9) and integrating with respect to the velocity leads to the advection-diffusion equation

$$\partial_t \rho(x, t) + \partial_x F_{\text{eq}}(\rho(x, t)) = \varepsilon\partial_x(\mu(\rho)\rho_x(x, t)), \tag{10}$$

where the diffusion coefficient $\mu(\rho)$ is given by

$$\begin{aligned}\mu_{\text{BGK}}(\rho) &= \int_0^1 v^2\partial_\rho M_f(v; \rho)\mathrm{d}v - \left(\int_0^1 v\partial_\rho M_f(v; \rho)\mathrm{d}v\right)^2 \\ &= \int_0^1 v^2\partial_\rho M_f(v; \rho)\mathrm{d}v - F'_{\text{eq}}(\rho)^2.\end{aligned} \tag{11}$$

If $\mu(\rho) < 0$, then the advection-diffusion equation is ill-posed and therefore may exhibit solutions with unbounded growth. In the case of the kinetic model (9), the sign of the diffusion coefficient depends on the equilibrium distribution M_f. The request $\mu_{\text{BGK}}(\rho) > 0$ is

$$\partial_\rho\left(\int_0^1 v^2 M_f(v; \rho)\mathrm{d}v\right) > F'_{\text{eq}}(\rho)^2 \tag{12}$$

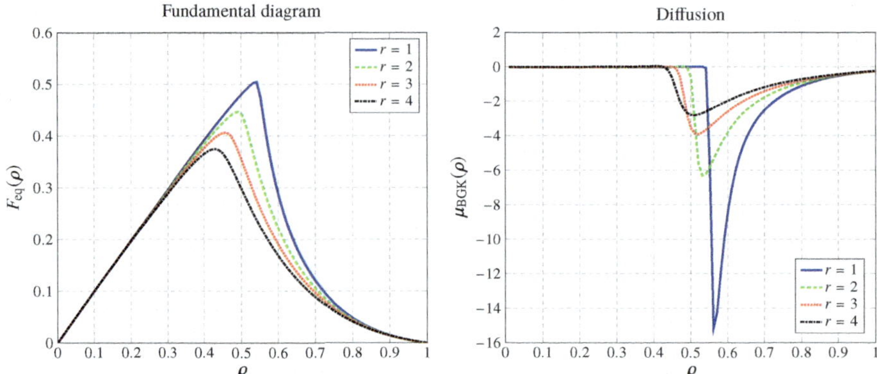

Fig. 6 The right panel shows the sign of the diffusion coefficient (11) for the BGK model (9) with the corresponding equilibrium distribution in the left panel

and, since $F_{eq}(\rho)$ is the fundamental diagram at equilibrium, this condition requires that the square of the characteristic velocities is bounded by the variation of the kinetic energy in each regime.

Below we recall the result in [11], which proves that the instability of the solution does not depend on the choice of the equilibrium distribution and in fact occurs for any suitable equilibrium of kinetic traffic models.

Proposition 1 *Assume that* $\exists\, \widetilde{\rho} \in (0, 1)$ *such that*

$$F'_{eq}(\rho) = \int_0^1 v\partial_\rho M_f(v;\rho)\mathrm{d}v < 0, \quad \partial_\rho \mathrm{Var}(v) = \partial_\rho \int_0^1 (v - U_{eq}(\rho))^2 M_f(v;\rho)\mathrm{d}v < 0 \tag{13}$$

for all $\rho \in (\widetilde{\rho}, 1)$. *Then the quantity* $\mu(\rho)$ *given in* (11) *is negative* $\forall\, \rho \in (\widetilde{\rho}, 1)$.

We analyze the validity of this result for the model in [19]. In Fig. 6 we investigate the sign of the diffusion coefficient (11) in the case of the equilibrium distribution. Those distributions are computed numerically for the spatially homogeneous kinetic model (1)–(3). Again, we use 48 discrete speeds, a fixed value of the acceleration parameter $\Delta_a = \frac{V_M}{4} = \frac{1}{4}$ and several values of the uncertainty Δ_b such that $r = \frac{\Delta_a}{\Delta_b} = 1, 2, 3, 4$. We observe that $\mu(\rho)_{\text{BGK}} \geq 0$ in the regime where the flux is increasing, while $\mu(\rho)_{\text{BGK}} < 0$ in the regime where the flux is decreasing. Increasing the uncertainty on the over-braking, the model becomes "less" unstable. In fact, the diffusion becomes larger but still negative. This may serve as explanation of the growth of perturbations in the density, numerically observed in the top of Fig. 5.

In view of the results provided by the Chapman-Enskog analysis, we state the following definition.

Definition 1 A mathematical model for traffic flow is said to be stable if its Chapman-Enskog expansion provides $\mu(\rho) \geq 0$, $\forall \rho \in [0, \rho_M]$, weakly unstable if $\mu(\rho) < 0$ on an interval (ρ_1, ρ_2) properly contained in $[0, \rho_M]$ and unstable if $\mu(\rho) < 0$ on an interval (ρ_1, ρ_2) in which either $\rho_1 = 0$ or $\rho_2 = \rho_M$.

The definition of a weakly unstable model is a consequence of the experimental observation in [11, 20] that if $\mu(\rho) < 0$ on an interval (ρ_1, ρ_2) properly contained in $[0, \rho_M]$, then the backward propagating waves in dense traffic remain bounded, because, when the oscillations reach $\rho = \rho_1$ and $\rho = \rho_2$, they fall in the diffusive region and they are damped. This leads to weak instabilities that in turn can be regarded as models for stop-and-go waves.

Concerning the concept of stability of Definition 1, the discrete BGK model for traffic introduced in [5] can be either stable or unstable. The model is characterized by non-local terms and with a suitable choice of the headway parameter the diffusion coefficient in the Chapman-Enskog expansion is positive on $[0, \rho_M]$. As observed in [20], this is not desirable in a model for traffic flow since it would not allow to reproduce non-equilibrium phenomena, such as stop-and-go waves.

3.2 *The Aw-Rascle and Zhang Model*

The Aw-Rascle and Zhang (ARZ) model will be considered in view of the stability analysis following [11]. We will show that it is weakly unstable. This justifies the derivation of a new BGK-type model in Sect. 4. The following result was already mentioned and analyzed in [20].

The ARZ model reads in primitive variables as

$$\begin{aligned} &\partial_t \rho(x,t) + \partial_x(\rho(x,t)u(x,t)) = 0 \\ &\partial_t\big(u(x,t) + h(\rho)\big) + u(x,t)\partial_x\big(u(x,t) + h(\rho)\big) = \frac{1}{\varepsilon}(U_{\text{eq}}(\rho) - u(x,t)), \end{aligned} \tag{14}$$

where u is the macroscopic speed of the flow and the function $h = h(\rho)$ is a strictly increasing function of the density and it is called hesitation function or traffic pressure. The quantity ε is a time which rules the relaxation speed of the velocity u to the equilibrium speed $U_{\text{eq}}(\rho)$ which is a given function of the density. Here U_{eq} is not necessarily given by (5).

System (14) can be understood as a relaxation system [12] converging towards the conservation law given by the Lighthill-Whitham [14] and Richards [18] model in the limit $\varepsilon \to 0$. If ε is small, but not vanishing, (14) approaches the advection-diffusion equation (10) where the diffusion coefficient $\mu(\rho)$ is given by

$$\mu_{\text{ARZ}}(\rho) = -\rho(x,t)^2 U'_{\text{eq}}(\rho)\big(U'_{\text{eq}}(\rho) + h'(\rho)\big). \tag{15}$$

This result is again obtained via Chapman-Enskog expansion, by considering a first-order expansion of the speed $u = U_{\text{eq}}(\rho) + \varepsilon u_1$ around the equilibrium velocity function $U_{\text{eq}}(\rho)$. The condition $\mu(\rho) > 0$ provides the so-called sub-characteristic condition [6, 12]. For the ARZ model $\mu(\rho) > 0$ is satisfied if

$$0 > U'_{\text{eq}}(\rho) > -h'(\rho). \tag{16}$$

We stress the fact that condition (16) strongly restricts the possible choice of U_{eq} and h, which can be chosen in order to make the model weakly unstable.

4 The Modified Formulation of the BGK Approximation in Traffic Flow

The derivation of the modified BGK-type equation for traffic flow is shortly summarized and we refer to [11] for a thorough discussion. The model is derived via mesoscopic limit of the microscopic follow-the-leader (FTL) and Bando model. We recall that the FTL-Bando model is proved to converge to the ARZ model in the macroscopic limit, both in one-dimension [1] and two-dimensions [10]. Therefore, the second-order system of moments of the new BGK model has also the property of representing a mesoscopic formulation of the class of second-order ARZ-type macroscopic models. As a consequence the feature of an ARZ-type model of having a negative diffusion coefficient in a small density regime is automatically obtained also for the new BGK-type equation.

4.1 BGK-Type Model Derived from the FTL-Bando Model

Let (x_i, v_i) be the microscopic states, position and velocity, of vehicle i. The follow-the-leader and Bando model is

$$\begin{aligned} \dot{x}_i &= v_i = w_i - p(\rho_i) \\ \dot{w}_i &= \frac{1}{\varepsilon}(U_{\text{eq}}(\rho_i) + p(\rho_i) - w_i), \end{aligned} \tag{17}$$

where $w_i := v_i + p(\rho_i)$, and the function $p = p(\rho_i)$ is the so-called traffic pressure. We assume that p satisfies $p(\rho) \geq 0$, $p'(\rho) > 0$ and

$$\frac{\mathrm{d}}{\mathrm{d}t} p(\rho_i) = -K(x_i, x_{i+1}, v_i, v_{i+1}),$$

where K is a term describing the interactions among vehicles. In the classical FTL model

$$K(x_i, x_{i+1}, v_i, v_{i+1}) = C_\gamma \frac{v_{i+1} - v_i}{(x_{i+1} - x_i)^{\gamma+1}},$$

where the constants $C_\gamma > 0$ and $\gamma > 0$ are given parameters. However, we consider the case of a general function K. The introduction of the quantity w_i allows us to rewrite the classical Bando model as a relaxation step (17).

Let now $g = g(x, w, t) : \mathbb{R} \times W \times \mathbb{R}^+ \to \mathbb{R}^+$ be the kinetic distribution function with respect to the desired speed w, which is assumed to be the speed that drivers want to keep in "optimal" situations. We define $W := [w_{\min}, +\infty)$ the space of the microscopic desired speeds where $w_{\min} > 0$ may be interpreted as the minimum speed limit in free flow conditions. The macroscopic density, i.e., the number of vehicles per unit length, at time t and position x is defined by

$$\rho(x, t) := \int_W g(x, w, t) \mathrm{d}w, \tag{18}$$

and we define the macroscopic quantity

$$q(x, t) := \int_W w g(x, w, t) \mathrm{d}w. \tag{19}$$

The derivation of the evolution equation for the kinetic distribution $g = g(x, w, t)$ is performed by reformulating the microscopic particle model (17) in a probabilistic interpretation and allowing a relaxation towards a desired distribution $M_g = M_g(w; \rho)$, as in [15, Section 4.2.2]. The distribution M_g has to fulfill the requirement

$$\int_W M_g(w; \rho) \mathrm{d}w = \rho(x, t),$$

and additionally

$$\frac{1}{\rho(x, t)} \int_W w M_g(w; \rho) \mathrm{d}w = U_{\mathrm{eq}}(\rho) + p(\rho). \tag{20}$$

According to [15, Section 4.2.2] and [11], it is possible to show that g solves

$$\partial_t g(x, w, t) + \partial_x\big[(w - p(\rho)) g(x, w, t)\big] = \frac{1}{\varepsilon} \big(M_g(w; \rho) - g(x, w, t)\big). \tag{21}$$

This equation is still a BGK-type equation since the collision kernel is linear and describes the relaxation of g towards a given distribution M_g parameterized by the density ρ. For a detailed derivation of (21) we refer to [11].

It is important to point out that, compared to classic kinetic theory, this approach is different in the sense that M_g is an "equilibrium distribution" with a modified microscopic velocity. Thanks to (20), M_g is imposed a-priori but it is still based on the knowledge of the classical Maxwellian M_f, which is related to the classical concept of microscopic velocity, by means of $U_{\text{eq}}(\rho) := \frac{1}{\rho}\int vM_f \mathrm{d}v$. In other words, M_f is not imposed a-priori (and so $U_{\text{eq}}(\rho)$ and consequently M_g), but the equilibrium distribution M_f is the one obtained by the modeling of microscopic interactions of the spatially homogeneous kinetic model. Any Maxwellian M_f of a kinetic model for traffic can be used to define M_g and the BGK model (21). Here, we study the Maxwellian M_f provided by [19].

4.2 Chapman-Enskog Expansion of the Modified BGK Model

We perform a Chapman-Enskog expansion for the model (21). We consider a first-order perturbation of g as

$$g(x, w, t) = M_g(w; \rho) + \varepsilon g_1(x, w, t), \quad \text{with} \int_W g_1(x, w, t)\mathrm{d}w = 0$$

and define $F_{\text{eq}}(\rho) = \rho U_{\text{eq}}(\rho)$. Then, it is possible to show, cf. [11], that the BGK-type equation (21) solves the advection-diffusion equation (10) with

$$\mu(\rho) = -F'_{\text{eq}}(\rho)^2 + \int_V v^2 \partial_\rho M_f(v; \rho)\mathrm{d}v - \rho p'(\rho)F'_{\text{eq}}(\rho) + F_{\text{eq}}(\rho)p'(\rho). \tag{22}$$

Observe that, compared to (11), the diffusion coefficient (22) contains two additional terms which depend on the function $p(\rho)$. Therefore, it is possible, for a given distribution M_f, to find a suitable $p(\rho)$ such that $\mu(\rho) > 0$ also in the congested regime. In particular, it is possible to find $p(\rho)$ in order to guarantee that the model is weakly unstable. Recall that $\mu_{\text{BGK}}(\rho)$ given in (11) was unconditionally negative in the congested phase of traffic for the classical BGK model (9).

Setting $F_{\text{eq}}(\rho) = \rho U_{\text{eq}}(\rho)$ the second two terms of the diffusion coefficient (22) can be written in terms of the equilibrium speed function as

$$\mu(\rho) = -F'_{\text{eq}}(\rho)^2 + \int_V v^2 \partial_\rho M_f(v; \rho)\mathrm{d}v - \rho^2 p'(\rho)U'_{\text{eq}}(\rho). \tag{23}$$

Therefore, $\mu(\rho) = \mu_{\text{BGK}}(\rho) + C(\rho)$ where $C(\rho) = -\rho^2 p'(\rho)U'_{\text{eq}}(\rho) \geq 0$ since p and U_{eq} are an increasing and a non-increasing function of the density, respectively. This means that, for $C(\rho)$ sufficiently large, the additional term yields a negative diffusion coefficient (11) in a bounded domain contained in $[0, \rho_M]$. In Fig. 7 we numerically show this result for the case of the homogeneous kinetic model in [19]. We consider the Maxwellian computed numerically with 48 discrete speeds, $\Delta_a =$

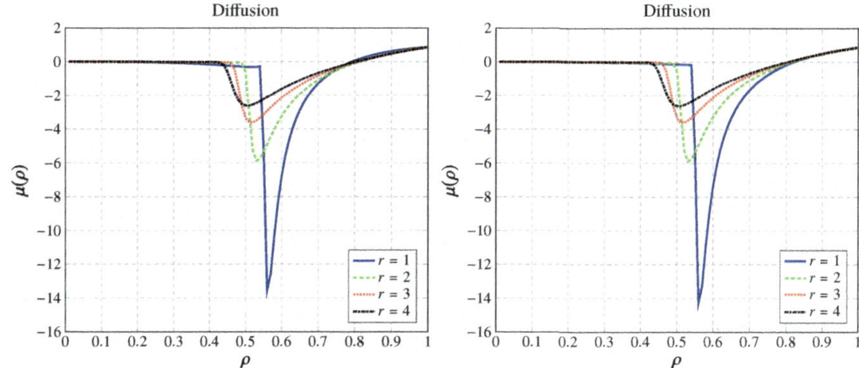

Fig. 7 Diffusion coefficient (23) for $p(\rho) = \frac{3}{2}\rho^2$ (left) and $p(\rho) = \rho^3$ (right)

$\frac{1}{4}$ and $\Delta_b = \frac{\Delta_a}{r}$, $r = 1, 2, 3, 4$. The pressure function is chosen as $p(\rho) = \frac{3}{2}\rho^2$ (left panel) and $p(\rho) = \rho^3$ (right panel).

5 Conclusions and Future Perspectives

In this work we have focused on the formulation of kinetic models for vehicular traffic flow which reproduce backward propagating waves in dense traffic. The underlying kinetic model is the one introduced in [19]. Backward traveling waves have been observed by defining an interaction rate that is a function of the density and its derivative.

A stability analysis of the waves in dense traffic regimes has been performed on the BGK-type approximation, in the limit of small interaction rates. We have shown that the model leads to an advection-diffusion equation with a negative diffusion coefficient in the whole congested regime, therefore producing an unbounded growth of dense waves in time. This justified to reconsider the results of [19] in the framework of a novel BGK formulation [11]. Finally, the formulation allows to have a weakly unstable model with results that show the existence of stop-and-go waves.

Acknowledgments The research of M. Herty and G. Visconti is funded by the Deutsche Forschungsgemeinschaft (DFG, German Research Foundation) under Germany's Excellence Strategy—EXC-2023 Internet of Production—390621612 as well as by DFG HE5386/13.

G. Puppo and G. Visconti acknowledge also support from GNCS (Gruppo Nazionale per il Calcolo Scientifico) of INdAM (Istituto Nazionale di Alta Matematica), Italy.

References

1. A. Aw, A. Klar, T. Materne, M. Rascle, Derivation of continuum traffic flow models from microscopic follow-the-leader models. SIAM J. Appl. Math. **63**(1), 259–278 (2002)
2. A. Aw, M. Rascle, Resurrection of "second order" models of traffic flow. SIAM J. Appl. Math. **60**(3), 916–938 (electronic) (2000)
3. M. Bando, K. Hasebe, A. Nakayama, A. Shibata, Y. Sugiyama, Dynamical model of traffic congestion and numerical simulation. Phys. Rev. E **51**(2), 1035–1042 (1995)
4. P.L. Bhatnagar, E.P. Gross, M. Krook, A model for collision processes in gases. I. Small amplitude processes in charged and neutral one-component systems. Phys. Rev. **94**(3), 511–525 (1954)
5. R. Borsche, A. Klar, A nonlinear discrete velocity relaxation model for traffic flow. SIAM J. Appl. Math. **78**(5), 2891–2917 (2018)
6. G.-q. Chen, C.D. Levermore, T.-P. Liu, Hyperbolic conservation laws with stiff relaxation terms and entropy. Comm. Pure Appl. Math. **47**, 787–830 (1992)
7. G. Dimarco, L. Pareschi, Asymptotic preserving implicit-explicit Runge-Kutta methods for non linear kinetic equations. SIAM J. Num. Anal. **51**, 1064–1087 (2013)
8. L. Fermo, A. Tosin, A fully-discrete-state kinetic theory approach to modeling vehicular traffic. SIAM J. Appl. Math. **73**(4), 1533–1556 (2013)
9. D. Gazis, R. Herman, R. Rothery, Nonlinear follow-the-leader models of traffic flow. Oper. Res. **9**(4), 545–567 (1961)
10. M. Herty, S. Moutari, G. Visconti, Macroscopic modeling of multilane motorways using a two-dimensional second-order model of traffic flow. SIAM J. Appl. Math. **78**(4), 2252–2278 (2018)
11. M. Herty, G. Puppo, S. Roncoroni, G. Visconti, The BGK approximation of kinetic models for traffic. Kinet. Relat. Models (2020, in press)
12. S. Jin, Z. Xin, The relaxation schemes for systems of conservation laws in arbitrary space dimensions. Comm. Pure Appl. Math. **48**, 235–277 (1995)
13. A. Klar, R. Wegener, Enskog-like kinetic models for vehicular traffic. J. Stat. Phys. **87**, 91 (1997)
14. M.J. Lighthill, G.B. Whitham, On kinematic waves. II. A theory of traffic flow on long crowded roads. Proc. Roy. Soc. Lond. Ser. A. **229**, 317–345 (1955)
15. L. Pareschi, G. Toscani, *Interacting Multiagent Systems. Kinetic equations and Monte Carlo methods* (Oxford University Press, 2013)
16. G. Puppo, M. Semplice, A. Tosin, G. Visconti, Analysis of a multi-population kinetic model for traffic flow. Commun. Math. Sci. **15**(2), 379–412 (2017)
17. G. Puppo, M. Semplice, A. Tosin, G. Visconti, Kinetic models for traffic flow resulting in a reduced space of microscopic velocities. Kinet. Relat. Mod. **10**(3), 823–854 (2017)
18. P.I. Richards, Shock waves on the highway. Oper. Res. **4**, 42–51 (1956)
19. S. Roncoroni, Kinetic modelling of vehicular traffic flow. Technical report, Università degli Studi dell'Insubria, 2017. Master Thesis.
20. B. Seibold, M.R. Flynn, A.R. Kasimov, R.R. Rosales, Constructing set-valued fundamental diagrams from Jamiton solutions in second order traffic models. Netw. Heterog. Media **8**(3), 745–772 (2013)
21. H.M. Zhang, A non-equilibrium traffic model devoid of gas-like behavior. Transp. Res. B-Meth. **36**(3), 275–290 (2002)

Structural Properties of the Stability of Jamitons

Rabie Ramadan, Rodolfo Ruben Rosales, and Benjamin Seibold

Abstract It is known that inhomogeneous second-order macroscopic traffic models can reproduce the phantom traffic jam phenomenon: whenever the sub-characteristic condition is violated, uniform traffic flow is unstable, and small perturbations grow into nonlinear traveling waves, called jamitons. In contrast, what is essentially unstudied is the question: which jamiton solutions are dynamically stable? To understand which stop-and-go traffic waves can arise through the dynamics of the model, this question is critical. This paper first presents a computational study demonstrating which types of jamitons do arise dynamically, and which do not. Then, a procedure is presented that characterizes the stability of jamitons. The study reveals that a critical component of this analysis is the proper treatment of the perturbations to the shocks, and of the neighborhood of the sonic points.

1 Introduction

The modeling of vehicular traffic flow via mathematical equations is a key building block in traffic simulation, state estimation, and control. Important ways to describe traffic flow dynamics are microscopic/vehicle-based [5, 47, 54], cellular [11, 46], and continuum models. This last class is the focus of this paper, particularly: inviscid macroscopic models [3, 37, 43, 51, 52, 55, 62] that describe the spatio-temporal evolution of the vehicle density (and other field quantities) via hyperbolic

R. Ramadan · B. Seibold (✉)
Department of Mathematics, Temple University, Philadelphia, PA, USA
e-mail: rabie.ramadan@temple.edu; seibold@temple.edu
https://www.math.temple.edu/~seibold

R. R. Rosales
Department of Mathematics, Massachusetts Institute of Technology, Cambridge, MA, USA
e-mail: rrr@math.mit.edu

G. Puppo, A. Tosin (eds.), *Mathematical Descriptions of Traffic Flow: Micro, Macro and Kinetic Models*, SEMA SIMAI Springer Series 12,
https://doi.org/10.1007/978-3-030-66560-9_3

conservation laws. Other types of continuum models exist as well, including gas-kinetic [26, 28, 53], dispersive [34, 35], and viscous [32, 33] models. Hyperbolic models do not resolve zones of strong braking, but rather approximate them by traveling discontinuities (shocks) whose dynamics are described by appropriate jump conditions [15]. Macroscopic models play a central role in traffic flow theory and practice because:

- Mathematically, other types of descriptions reduce/converge to macroscopic models in certain limits, including: microscopic [4], cellular [1], and gas-kinetic [1, 28].
- Practically, macroscopic models are best-suited for state estimation [63, 66], for incorporating sparse GPS data [2, 27], and for control [50].
- Computationally, a macroscopic description is a natural framework to upscale millions of vehicles to a cell-transmission model [11] with much fewer degrees of freedom.
- Societally, traffic descriptions that do not resolve individual vehicles are desirable for privacy and data security.

In this work, we focus on the lane-aggregated description of traffic flow dynamics on uniform highways without any road variations, let alone intersections or bottlenecks. The reason is that even in this simple scenario, real traffic flow tends to develop complex nonlinear dynamics, particularly the *phantom traffic jam* phenomenon [24, 31]: initially uniform flow develops (under small perturbations) into nonlinear traveling waves, called *jamitons* [19]. This occurrence of instabilities and waves without discernible reason has been demonstrated and reproduced experimentally [57, 59]. While these features can be reproduced in microscopic car-following models, a key goal is to capture these non-equilibrium phenomena via macroscopic models (to facilitate the model advantages described above).

The archetype macroscopic model is the *Lighthill-Whitham-Richards (LWR) model* [43, 55]

$$\rho_t + Q(\rho)_x = 0 \tag{1}$$

that describes the evolution of the vehicle density $\rho(x, t)$ where x is the road position and t is time. The *fundamental diagram* (FD) function $Q(\rho) = \rho U(\rho)$, where the *equilibrium velocity function* $U(\rho)$ is the bulk flow velocity as a function of density, is motivated by the 1935 measurements by Greenshields [22], and many types of FD have been proposed [11, 20, 43, 48, 62]. As a matter of fact, real FD data exhibits a substantial spread in the congested regime [31]. More complex traffic models capture this spread [9, 16, 17, 56], but the LWR model does not. Yet, due to its simplicity it nevertheless is widely used. Moreover, as we highlight below, it also is motivated as a reduced equation for more complex models.

Another critical shortcoming of the LWR model is that it cannot reproduce the phantom traffic jam phenomenon: being a *first-order model*, it exhibits a maximum principle, and thus small perturbations to a uniform solution cannot amplify (instead, they turn into N-waves and decay). In this work, we focus on *second-order models* that augment the vehicle density $\rho(x,t)$ by an independent field variable for the bulk velocity $u(x,t)$, and describe their evolution via a 2×2 balance law system, specifically: a hyperbolic conservation law system with a relaxation term in the velocity equation. Due to conservation of vehicles, the density always evolves by the continuity equation, $\rho_t + (\rho u)_x = 0$. In turn, the velocity equation encodes the actual modeling of the vehicle dynamics and interactions. The *Payne-Whitham (PW) model* [51, 65]

$$\begin{aligned} \rho_t + (\rho u)_x &= 0\,, \\ u_t + u u_x + p(\rho)_x/\rho &= \tfrac{1}{\tau}(U(\rho) - u) \end{aligned} \tag{2}$$

was the first second-order model proposed. Here $U(\rho)$ is the *desired velocity function*, and τ is the relaxation time that determines how fast drivers adjust to their desired velocity $U(\rho)$. The *traffic pressure* $p(\rho)$ models preventive driving. Even though the PW model does capture traffic waves accurately [19, 56], it is generally rejected [12] due to spurious shocks that overtake vehicles from behind; and other hyperbolic models are preferred (see below). However, the fundamental structure of a 2×2 hyperbolic system with a relaxation in the second equation is common to all models of interest in this study.

Models with the structure described above possess a critical phase transition. If the *sub-characteristic condition* (SCC) is satisfied, then uniform flow is stable [7, 44, 64, 65]. Conversely, when it is violated, uniform flow is unstable and nonlinear traveling wave solutions exist [19, 29, 41, 49, 56]. The SCC is defined as follows. Let $\lambda_1 < \lambda_2$ be the two characteristic speeds of the hyperbolic part of the model, and let $\mu = Q'(\rho)$ be the characteristic speed of the *reduced equation* (1) (with $Q(\rho) = \rho U(\rho)$), which arises in the formal limit $\tau \to 0$; in which u relaxes infinitely fast to $U(\rho)$. Then the SCC is: $\lambda_1 \le \mu \le \lambda_2$.

The case of the SCC satisfied is well studied [7, 42, 44, 64, 65]. In particular, it is related to positive diffusion when conducting a Chapman-Enskog expansion of the model [25, 35]. In contrast, this paper focuses on understanding the behavior and stability of solutions when the SCC is violated.

This paper is organized as follows. In Sect. 2, we introduce the equations. Then we characterize the nature of the instabilities to uniform flow, and the traveling wave solutions that then arise: the jamitons. In Sect. 3, a systematic computational study of the stability of jamitons is conducted. Those results then motivate a stability analysis of those nonlinear traveling waves, presented in Sect. 4. We close with a discussion and a broader outlook in Sect. 5.

2 Macroscopic Traffic Models with Instabilities and Traveling Waves

While the general results and methodologies apply to a wide class of second-order models with relaxation (including the PW model (2) and generic second-order models [16, 38]), we focus this study on the inhomogeneous *Aw-Rascle-Zhang (ARZ) model* [3, 67]. In non-conservative form it reads as

$$\begin{aligned} \rho_t + (\rho u)_x &= 0\,, \\ (u + h(\rho))_t + u(u + h(\rho))_x &= \tfrac{1}{\tau}(U(\rho) - u)\,, \end{aligned} \tag{3}$$

where $h(\rho)$ is called the *hesitation function*. We assume that: $U(\rho)$ is strictly decreasing, $Q(\rho) = \rho U(\rho)$ is strictly concave, $h(\rho)$ is strictly increasing, and $\rho h(\rho)$ is strictly convex. In particular these assumptions yield a hyperbolic system, which has no waves that overtake vehicles (the 2-waves are contacts) [3]. While originally proposed in homogeneous form, the addition of the relaxation term [21] allows for the violation of the SCC.

In the homogeneous ARZ model, the field $w = u + h(\rho)$ can be interpreted as a convected quantity moving with the flow (the hesitation function reduces the empty road velocity w by $h(\rho)$). Hence, the conserved variables are ρ and $q = \rho(u+h(\rho))$, and the conservative form of the equations is

$$\begin{aligned} \rho_t + (q - \rho h(\rho))_x &= 0\,, \\ q_t + \left(\tfrac{q^2}{\rho} - q h(\rho)\right)_x &= \tfrac{1}{\tau}\left(\rho(U(\rho) + h(\rho)) - q\right)\,, \end{aligned} \tag{4}$$

with associated Rankine-Hugoniot jump conditions

$$\begin{aligned} s\,[\rho] - [\rho u] &= 0\,, \\ s\left[\rho\big(u + h(\rho)\big)\right] - \left[\rho u^2 + \rho u h(\rho)\right] &= 0\,. \end{aligned} \tag{5}$$

Here $[\zeta]$ denotes the jump of the variable ζ across the discontinuity, and s is the speed. In addition, the Lax entropy conditions [15] must be satisfied. Specifically: one family of characteristics goes through the discontinuity, while the other converges into it (for a shock), or is parallel to it (for a contact). In particular, the assumptions on h made below (3) guarantee that the entropy conditions are equivalent to: the shocks are compressive (i.e., as vehicles go through a shock, the density increases) and move slower than the vehicles [56].

The characteristic speeds of (4) are

$$\lambda_1 = q/\rho - h(\rho) - \rho h'(\rho) = u - \rho h'(\rho)\,, \quad \text{and} \quad \lambda_2 = q/\rho - h(\rho) = u\,, \tag{6}$$

where the λ_1 is genuinely nonlinear (associated with shocks and rarefactions), while the λ_2 is linearly degenerate (associated with contacts).

2.1 Specific Model Functions

While the analysis and general results derived below hold for generic models (4), the computational study and the illustrative graphs are presented for a specific choice of model functions. As in [56], we choose $\rho_{\max} = 1/7.5\text{m}$, $u_{\max} = 20\text{m/s}$, and construct the fundamental diagram function

$$Q(\rho) = c\left(g(0) + (g(1) - g(0))\,\frac{\rho}{\rho_{\max}} - g\left(\frac{\rho}{\rho_{\max}}\right)\right), \quad \text{where } g(y) = \sqrt{1 + \left(\tfrac{y-b}{\lambda}\right)^2},$$

that is a smoothed version of the Newell-Daganzo triangular flux [11, 48]. The parameters are chosen $c = 0.078\rho_{\max}u_{\max}$, $b = \frac{1}{3}$, and $\lambda = \frac{1}{10}$ to have the function fit real sensor data [56]. Hence $U(\rho) = Q(\rho)/\rho$. Moreover, we choose $h(\rho) = 8\text{m/s}\sqrt{\frac{\rho}{\rho_{\max}-\rho}}$, and the relaxation time $\tau = 3\text{s}$. Note that these values are for a single lane. When considering multi-lane traffic, realistic values result by scaling ρ and Q by the number of lanes.

2.2 Linear Stability of Uniform Flow

Before analyzing the stability of nonlinear waves, we discuss important aspects regarding the stability of uniform flow, i.e., base state solutions of (3) in which $\rho = \tilde{\rho}$ and $u = U(\tilde{\rho})$ are constant in space and time. The linear stability analysis itself is a well-established normal model analysis [19, 32], and we briefly outline the key steps. Consider infinitesimal wave perturbations (where k is the wave number and σ the complex growth rate) of the base state ,

$$\hat{\rho} = \hat{R}e^{ikx+\sigma t} \quad \text{and} \quad \hat{u} = \hat{U}e^{ikx+\sigma t},$$

substitute the perturbed solution $\rho = \tilde{\rho} + \hat{\rho}$ and $u = U(\tilde{\rho}) + \hat{u}$ into (3), and consider only constant and linear terms. This leads to the system

$$\begin{bmatrix} \sigma + ik\psi & ik\tilde{\rho} \\ \sigma\phi + ik\psi\phi - \frac{\xi}{\tau} & \sigma + ik\psi + \frac{1}{\tau} \end{bmatrix} \begin{bmatrix} \hat{R} \\ \hat{U} \end{bmatrix} = \begin{bmatrix} 0 \\ 0 \end{bmatrix}, \tag{7}$$

for the perturbation amplitudes, where $\psi = U(\tilde{\rho}) > 0$, $\phi = h'(\tilde{\rho}) > 0$, and $\xi = U'(\tilde{\rho}) < 0$. Nontrivial solutions can only exist if the matrix in (7) has vanishing determinant, which requires

$$\sigma = -ik\psi + ik\tfrac{1}{2}\tilde{\rho}\phi - \tfrac{1}{2\tau}(1+\Gamma)\,,$$

where Γ satisfies $\Gamma^2 = 1 - k^2\tau^2\tilde{\rho}^2\phi^2 - 2ik\tau\tilde{\rho}(\phi + 2\xi)$. Writing $\Gamma = \Lambda_1 + i\Lambda_2$ in terms of its real and imaginary part yields the two equations $\Lambda_1^2 - \Lambda_2^2 = 1 - k^2\tau^2\tilde{\rho}^2\phi^2$ and $\Lambda_1\Lambda_2 = k\tau\tilde{\rho}(\phi + 2\xi)$, which then leads to the following quadratic equations for $z = (\Lambda_1)^2$:

$$z^2 - (1 - \beta^2k^2)z - \gamma^2k^2 = 0\,. \tag{8}$$

Here $\beta = \tau\tilde{\rho}\phi$ and $\gamma = \tau\tilde{\rho}(\phi + 2\xi)$. The positive solution of (8), as a function of k, is

$$z^+(k) = \tfrac{1}{2}\left((1-\beta^2k^2) + \sqrt{(1-\beta^2k^2)^2 + 4\gamma^2k^2}\right) \tag{9}$$

$$= \tfrac{1}{2}\left((1-\beta^2k^2) + \sqrt{(1+\beta^2k^2)^2 + 4(\gamma^2-\beta^2)k^2}\right)\,. \tag{10}$$

This function has the following properties:

(i) $z^+(0) = 1$.

(ii) $\lim_{k\to\infty} z^+(k) = (\gamma/\beta)^2$, which follows from (9) and the asymptotic ($k \gg 1$) formula:

$$\sqrt{(1-\beta^2k^2)^2 + 4\gamma^2k^2} \sim \beta^2k^2\sqrt{1 + 2(2\gamma^2 - \beta^2)\beta^{-4}k^{-2}} \sim \beta^2k^2 + (2(\gamma/\beta)^2 - 1)\,.$$

(iii) It is strictly monotonic if $|\gamma| \neq |\beta|$, i.e., it is strictly increasing if $|\gamma| > |\beta|$ and strictly decreasing if $|\gamma| < |\beta|$. This fact follows from (9), because the sign of the term $4(\gamma^2 - \beta^2)k^2$ determines the slope of $z^+(k)$: if $|\gamma| = |\beta|$, it is constant; and if the term is positive (negative), the function goes up (down) with k.

The growth rate of normal modes is

$$g_{\tilde{\rho}}(k) = \mathrm{Re}(\sigma) = -\tfrac{1}{2\tau}(1 + \mathrm{Re}(\Gamma)) = -\tfrac{1}{2\tau}(1 + \Lambda_1) = -\tfrac{1}{2\tau}\left(1 \pm \sqrt{z^+(k)}\right)\,.$$

Linear stability, i.e., $\mathrm{Re}(\sigma) \leq 0$, is equivalent to $z^+ \leq 1$ (only the negative root of $\sqrt{z^+}$ could cause positive growth). Hence, stability holds exactly if $|\gamma| < |\beta|$, or equivalently $\phi + \xi > 0$, or equivalently

$$h'(\tilde{\rho}) + U'(\tilde{\rho}) \geq 0\,. \tag{11}$$

This last condition is exactly what the sub-characteristic condition (SCC) [64, 65] yields as well [56]: the LWR characteristic speed, $\mu = Q'(\tilde{\rho}) = U(\tilde{\rho}) + \tilde{\rho}U'(\tilde{\rho})$ lies in between the two ARZ characteristic speeds, $\lambda_1 = U(\tilde{\rho}) - \tilde{\rho}h'(\tilde{\rho})$ and $\lambda_2 = U(\tilde{\rho})$, exactly if (11) holds.

To recap, for the inhomogeneous ARZ model (3), there are exactly two possibilities: Either the stability condition (the SCC) (11) holds; then all basic wave perturbations e^{ikx} have non-positive growth rates, and solutions are linearly stable. Or (11) is violated; then all waves grow. Moreover, the rate of growth $g_{\tilde{\rho}}(k)$ is an increasing function of the wave number k, that has $g_{\tilde{\rho}}(0) = 0$, and approaches (as $k \to \infty$) the asymptotic growth rate

$$g_{\tilde{\rho}}^{\infty} = \lim_{k\to\infty} g_{\tilde{\rho}}(k) = \tfrac{1}{2\tau}(|\gamma/\beta| - 1) = \tfrac{1}{2\tau}(|1 + 2\xi/\phi| - 1) = \tfrac{1}{\tau}\left(\tfrac{-U'(\tilde{\rho})}{h'(\tilde{\rho})} - 1\right).$$

Figure 1 shows the growth rate functions $g_{\tilde{\rho}}(k)$ for the specific model given in Sect. 2.1, with stable base states in Fig. 1a and unstable base states in Fig. 1b. In the latter, one can clearly see the strict increase of $g_{\tilde{\rho}}$ with k, and the asymptotic limit $g_{\tilde{\rho}}^{\infty}$. Figure 1c shows a plot of the asymptotic growth rate $g_{\tilde{\rho}}^{\infty}$ as a function of $\tilde{\rho}$.

Clearly, base states that satisfy (11) are well-behaved. However, with regards to modeling phantom traffic jams and jamitons, we are particularly interested in base states that violate (11). These require some more careful discussion. While instabilities to uniform states are ubiquitous in science and engineering, having a growth rate that is increasing for all wave numbers is unusual. The much more common scenario (for example, fluid instabilities moderated by viscosity or surface tension [13]) is that medium wave length are unstable and short waves (i.e., k large) are stable again, yielding a critical wave number k^* of maximal growth. In that case, one can argue that out of infinitesimal perturbations, in which all wave lengths are present, the linearized dynamics will single out the ones with dominant growth. Hence, the wave number k^* will be selected to first enter the nonlinear regime.

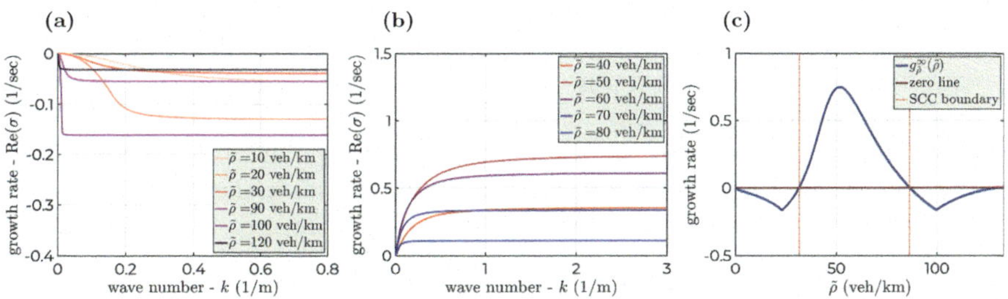

Fig. 1 Plots of the growth rate $g_{\tilde{\rho}}(k) = \mathrm{Re}(\sigma)$ as a function of the wave number k, for different constant base states $\tilde{\rho}$, as well as the asymptotic growth rate $g_{\tilde{\rho}}^{\infty}$ as a function of $\tilde{\rho}$. (**a**) Growth rates $g_{\tilde{\rho}}(k)$ for different $\tilde{\rho}$ that satisfy (11), i.e., are linearly stable. (**b**) Growth rates $g_{\tilde{\rho}}(k)$ for different $\tilde{\rho}$ that violate (11), i.e., are linearly unstable. (**c**) Asymptotic growth rate (worst case) $g_{\tilde{\rho}}^{\infty} = \lim_{k\to\infty} g_{\tilde{\rho}}(k)$ as a function of $\tilde{\rho}$

However, arguments of that type do not work for (3) because, as we have shown, its growth function $g_{\tilde{\rho}}(k)$ has no maximum. Rather, the shorter the waves in the perturbation, the faster their growth. It should be stressed that despite this behavior, the linearized model for (3) is mathematically well-posed: for any final time t, the amplification of normal modes is bounded by $\exp(t\, g_{\tilde{\rho}}^{\infty})$. Still, from an application perspective, properly answering the question of which wave lengths dominate once an amplified perturbation leaves the linear regime, is important; but it is more challenging than in the usual situation.

While the PDE model (3) has no maximum wave number, reality does, namely the vehicle scale. Specifically, wave numbers beyond a $k_{\max}$, given by the minimum spacing between vehicles, have no practical meaning. One possible way to exclude features on such unphysically short length scales is to add a small amount of viscosity to the ARZ model (3), as in Kerner-Konhäuser [32, 33] for the PW model (2). In Fig. 1b, this would change the functions $g_{\tilde{\rho}}(k)$ to drop off once k gets close to the vehicle scale. Similarly, the numerical discretization of the PDE (3) on grids that are never finer than the vehicle scale will produce a wave number cut-off via numerical viscosity of the method [39].

Another possibility (employed here in Sect. 3) is to consider small perturbations, rather than infinitesimal perturbations, and provide a model for the noise. Specifically, we argue that on real roads, perturbations of all wave lengths $k \in [0, k_{\max}]$ will act: $k < k_{\max}$ due to small variations in road features, wind, etc.; and $k \approx k_{\max}$ due to variabilities across vehicles. The simplest such noise model is one where all wave numbers $k \in [0, k_{\max}]$ appear with equal amplitudes, and perturbations with $k > k_{\max}$ do not occur.

Because the growth function tends to have a plateau near $k_{\max}$ (see Fig. 1), this linear growth/noise model will yield that all wave numbers k near but below $k_{\max}$ will be amplified to reach the nonlinear regime at the same time. This is not unrealistic, as it means that noise close to the vehicle scale will dominate before systematic nonlinear wave effects kick in.

As a final remark we wish to point out that once solutions of the ARZ model (3) leave the linear regime (around a uniform base state), the nonlinear dynamics tend to turn those vehicle-scale waves into oscillations with shocks that then collide and merge to form nonlinear wave structures of much smaller amplitude to wave-length ratios. However, those nonlinear transient dynamics are extremely complicated, and this insight is merely based on our observations from numerous highly resolved computations (like those done in Sect. 3). What we *will* study, though, is the stability of true traveling wave solutions of (3) (jamitons) in the situation when the SCC (11) is violated (see Sect. 4).

2.3 Traveling Wave Analysis and Jamitons

Before studying waves, it is important to stress that macroscopic models (without explicit lane changing) can equivalently be written in Lagrangian variables. In (4) the equations are cast in Eulerian variables $\rho(x,t)$ and $q(x,t)$. The Lagrangian formulation, as used in [21, 56], employs the variables $\mathrm{v}(\sigma,t)$ and $u(\sigma,t)$, where σ is the (continuous) vehicle number, defined so that $\mathrm{d}\sigma = \rho\,\mathrm{d}x - \rho u\,\mathrm{d}t$, and $\mathrm{v} = 1/\rho$ is the specific traffic volume, i.e., the road length per vehicle. In these variables the ARZ model reads as

$$\begin{aligned} \mathrm{v}_t - u_\sigma &= 0\ , \\ (u + \hat{h}(\mathrm{v}))_t &= \tfrac{1}{\tau}(\hat{U}(\mathrm{v}) - u)\ , \end{aligned} \tag{12}$$

where $\hat{h}(\mathrm{v}) = h(1/\mathrm{v})$ and $\hat{U}(\mathrm{v}) = U(1/\mathrm{v})$. The assumptions on the model functions in Eulerian variables ($\frac{\mathrm{d}U}{\mathrm{d}\rho} < 0$, $\frac{\mathrm{d}^2 Q}{\mathrm{d}\rho^2} < 0$, $\frac{\mathrm{d}h}{\mathrm{d}\rho} > 0$, $\frac{\mathrm{d}^2}{\mathrm{d}\rho^2}\rho h(\rho) > 0$) translate to the following assumptions in Lagrangian variables: $\frac{\mathrm{d}\hat{U}}{\mathrm{d}\mathrm{v}} > 0$, $\frac{\mathrm{d}^2\hat{U}}{\mathrm{d}\mathrm{v}^2} < 0$, $\frac{\mathrm{d}\hat{h}}{\mathrm{d}\mathrm{v}} < 0$, and $\frac{\mathrm{d}^2\hat{h}}{\mathrm{d}\mathrm{v}^2} > 0$. For simplicity, we now omit the hats, unless explicitly required for clarity. The characteristic speeds of (12) are $\lambda_1 = h'(\mathrm{v})$ and $\lambda_2 = 0$, and the associated Rankine-Hugoniot shock jump conditions are

$$\begin{aligned} m\,[\mathrm{v}] - [u] &= 0\ , \\ [u] + [h(\mathrm{v})] &= 0\ , \end{aligned} \tag{13}$$

where $-m$ is the propagation speed of the shock in the Lagrangian variables (in the Eulerian frame m is the flux of vehicles through the shock). Note that, for contact discontinuities, the conditions are: $m = 0$ and $[u] = 0$.

Below, we are going to employ both types of (equivalent) descriptions of the ARZ model. Eulerian (4) for the computational study of the nonlinear model in Sect. 3, and Lagrangian (12) for the jamiton stability analysis in Sect. 4.

Jamiton solutions can now be constructed via the Zel'dovich-von Neumann-Döring (ZND) theory [18]. One starts out with a traveling wave ansatz. In Eulerian variables, one seeks for solutions $\rho(x,t) = \rho(\eta)$, $u(x,t) = u(\eta)$ of (4) that depend on the single variable $\eta = \frac{x-st}{\tau}$. In Lagrangian variables, one considers solutions $\mathrm{v}(\sigma,t) = \mathrm{v}(\chi)$, $u(\sigma,t) = u(\chi)$ of (12), where $\chi = \frac{\sigma+mt}{\tau}$. Here s is the traveling wave speed in the road frame, while the Lagrangian wave speed $-m$ relates to the mass flux m of vehicles through the wave.

We start with the Lagrangian formulation [56]. The traveling wave ansatz leads to

$$\tfrac{m}{\tau}\mathrm{v}'(\chi) - \tfrac{1}{\tau}u'(\chi) = 0\ , \tag{14}$$

$$\tfrac{m}{\tau}u'(\chi) + h'(\mathrm{v}(\chi))\tfrac{m}{\tau}\mathrm{v}'(\chi) = \tfrac{1}{\tau}\left(U(\mathrm{v}(\chi)) - u(\chi)\right)\ . \tag{15}$$

Equation (14) yields that

$$m\mathrm{v} - u = -s\ , \tag{16}$$

where s is a constant of integration. Using (16) to substitute u by v in (15), we obtain the scalar first-order *jamiton ODE*

$$\mathrm{v}'(\chi) = \frac{w(\mathrm{v}(\chi))}{r'(\mathrm{v}(\chi))}, \tag{17}$$

where the two functions w and r are defined as

$$w(\mathrm{v}) = U(\mathrm{v}) - (m\mathrm{v} + s) \quad \text{and} \quad r(\mathrm{v}) = mh(\mathrm{v}) + m^2\mathrm{v}.$$

Because $h'(\mathrm{v}) < 0$ and $h''(\mathrm{v}) > 0$, the denominator in (17) has exactly one root, the *sonic value* $\mathrm{v_S}$ (occurring at the *sonic point*), such that $h'(\mathrm{v_S}) = -m$. The ODE (17) can be integrated through $\mathrm{v_S}$ if the numerator in (17) has a simple root at $\mathrm{v_S}$ as well. This leads to the *Chapman-Jouguet condition* [18]

$$m\mathrm{v_S} + s = U(\mathrm{v_S}),$$

which yields a relationship between the constants m and s as follows:

$$m = -h'(\mathrm{v_S}) \quad \text{and} \quad s = U(\mathrm{v_S}) - m\mathrm{v_S}.$$

One therefore has a one-parameter family of smooth traveling wave solutions, parameterized by $\mathrm{v_S}$, each being solutions of (17).

Into these smooth profiles shocks can be inserted that move with the same speed $-m$. The first condition in (13) implies that the quantity $m\mathrm{v} - u$ is conserved across the shock (in addition to being conserved along the smooth parts by (16)). And both conditions in (13) together imply that $r(\mathrm{v})$ is conserved across shocks. Hence, when integrating (17), one can at any value v^- insert a shock that jumps to a value v^+ with $r(\mathrm{v}^+) = r(\mathrm{v}^-)$ and continue integrating (17) from there. Moreover, one can only jump downwards to satisfy the Lax entropy conditions [65], i.e., $\mathrm{v}^+ < \mathrm{v_S} < \mathrm{v}^-$. This, in turn requires that the smooth jamiton profile $\mathrm{v}(\chi)$ must be an increasing function. Using L'Hôpital's rule in (17) at the sonic point yields that

$$0 < \frac{w'(\mathrm{v_S})}{r''(\mathrm{v_S})} = \frac{U'(\mathrm{v_S}) - m}{mh''(\mathrm{v_S})} = \frac{U'(\mathrm{v_S}) + h'(\mathrm{v_S})}{mh''(\mathrm{v_S})},$$

which means exactly that the SCC is violated. In other words, as shown in [56], jamiton profiles with shocks can exist if and only if the SCC is violated.

The construction in Eulerian variables is analogous, albeit a bit more technical (cf. [19]). The traveling wave ansatz leads to

$$-s\rho' + (\rho u)' = 0,$$
$$(u - s - \rho h'(\rho))u' = U(\rho) - u.$$

Integrating the first equation yields $\rho(u-s)=m$, which allows one to substitute ρ via u and vice versa. The second equation becomes the jamiton ODE

$$u'(\eta)=\frac{(u-s)(U(\rho)-u)}{(u-s)^2-mh'(\rho)}\,,$$

where $\rho=\frac{m}{u-s}$. The Chapman-Jouguet condition (matching roots of numerator and denominator) leads to the relations: $m=\rho_S^2h'(\rho_S)$ and $s=U(\rho_S)-\rho_S h'(\rho_S)$. Shock and entropy conditions are then implemented analogous to the Lagrangian situation.

With these rules, jamiton solutions can be constructed (in either choice of variables). For a given choice of v_S (and thus uniform propagation speed), any pattern of solutions to (17) connected by shocks (satisfying the above conditions) results in a traveling wave solution. The jamitons between any two shocks can be arbitrarily short (with a small variation around v_S), or may be arbitrarily long. In fact, it is not even required for the jamitons between shocks to have the same length (see [19, 56] for visualizations of jamiton profiles).

While all of these constitute feasible traveling wave solutions of the ARZ model (4), it does not mean that all such profiles would be dynamically stable under perturbations. In fact, both numerical evidence (see Sect. 3) as well as intuition dictate that neither very short nor very long jamitons should be stable. The former because they can be thought of as a small (sawtooth) perturbation of the constant v_S state (which is unstable because the SCC is violated, see Sect. 2.2); and the latter because their long tail will itself be close to a constant which, if that state violates the SCC, will be dynamically unstable. In other words, too short jamitons merge and have longer wave forms between them; and long jamitons have new instabilities grow in their tails. It is only the middle range of jamitons (not too short and not too long) that is expected to be dynamically stable; and only those should arise in actual practice.

This dynamic stability (of the jamitons themselves) has not been studied before. We do so, by first conducting a computational study in Sect. 3 that confirms the intuition above and quantifies it; and then deriving and analyzing linear perturbation equations for the jamiton solutions in Sect. 4.

3 Computational Study of Jamiton Stability

To understand the dynamic stability of jamitons, we conduct a systematic study of the ARZ model (4) via direct numerical computation. After constructing a periodic jamiton as outlined in Sect. 2.3, we insert that profile as an initial condition into a numerical scheme (Sect. 3.1) and investigate whether the profile is maintained under small perturbations (Sect. 3.2).

3.1 Numerical Scheme for the ARZ Model with Relaxation Term

The ARZ model (4) is a system of hyperbolic conservation laws with a relaxation term. The hyperbolic part of the system can be solved using a finite volume scheme based on an approximate Riemann solver [40]. To find the numerical flux at the cell boundaries, we use the HLL approximate Riemann solver [23], which guarantees that the numerical fluxes satisfy the entropy condition [36]. Given the grid cell $C_i = [x_i - \Delta x/2, x_i + \Delta x/2]$, where Δx is the cell size, let

$$U_i^n = \begin{bmatrix} \rho_i^n \\ q_i^n \end{bmatrix} \quad \text{and} \quad F_{i+\frac{1}{2}}^n = \begin{bmatrix} (F_\rho)_{i+\frac{1}{2}}^n \\ (F_q)_{i+\frac{1}{2}}^n \end{bmatrix}$$

denote the approximate solution (cell average) in cell C_i and the numerical flux at the boundary between cells C_i and C_{i+1}, respectively, at time $n\Delta t$ (n-th time step).

A numerically robust treatment of the relaxation term is achieved by treating it implicitly, resulting in the semi-implicit update rule

$$\begin{bmatrix} \rho_i^{n+1} \\ q_i^{n+1} \end{bmatrix} = \begin{bmatrix} \rho_i^n \\ q_i^n \end{bmatrix} - \tfrac{\Delta t}{\Delta x} \left(\begin{bmatrix} (F_\rho)_{i+\frac{1}{2}}^n \\ (F_q)_{i+\frac{1}{2}}^n \end{bmatrix} - \begin{bmatrix} (F_\rho)_{i-\frac{1}{2}}^n \\ (F_q)_{i-\frac{1}{2}}^n \end{bmatrix} \right) + \tfrac{\Delta t}{\tau} \begin{bmatrix} 0 \\ \rho_i^{n+1}\left(U(\rho_i^{n+1}) + h(\rho_i^{n+1})\right) - q_i^{n+1} \end{bmatrix}.$$

This ensures stability even when τ is small. Note that, because the implicit term appears only in the q-equation and because it is linear in q_i^{n+1}, the formally semi-implicit numerical scheme is actually fully explicit and the update step can be conducted in two sub-steps:

(1) Update the ρ component explicitly:

$$\rho_i^{n+1} = \rho_i^n - \tfrac{\Delta t}{\Delta x} \left((F_\rho)_{i+\frac{1}{2}}^n - (F_\rho)_{i-\frac{1}{2}}^n \right).$$

(2) Now, with ρ_i^{n+1} known from the first step, update

$$\left(1 - \tfrac{\Delta t}{\tau}\right) q_i^{n+1} = q_i^n - \tfrac{\Delta t}{\Delta x} \left((F_q)_{i+\frac{1}{2}}^n - (F_q)_{i-\frac{1}{2}}^n \right) + \tfrac{\Delta t}{\tau} \rho_i^{n+1} \left(U(\rho_i^{n+1}) + h(\rho_i^{n+1}) \right).$$

3.2 Results on the Stability of Jamitons

Using the numerical scheme described above, we conduct a computational investigation of the stability of jamitons (of the ARZ model (4) with the specific model functions and parameters described in Sect. 2.1). Specifically, we classify the jamitons as follows: Evolve the solution up to some large final time, while regularly adding small perturbations. Then a jamiton is classified as stable if the jamiton profile is (within a tolerance) maintained at the final time, and unstable otherwise.

To classify a given jamiton $J_0 = [\rho_0(x), u_0(x)]^T$ (of length L_0, with sonic density ρ_{s_0}, upstream density ρ_0^+, and speed s_0), we set up a periodic domain of length $4L_0$ with initial conditions $[\rho_{\text{ic}}(x), u_{\text{ic}}(x)]^T = [\rho_0(x \mod L_0), u_0(x \mod L_0)]^T$, i.e., the initial profile is four consecutive jamitons J_0 with shocks in between. We discretize using 10,000 grid cells, and run the numerical scheme (from Sect. 3.1) up to $t_{\text{final}} = 3{,}000$ (seconds; we omit units below).

During the numerical solution process, a small smooth perturbation is added to the vehicle velocity field $u = q/\rho - h(\rho)$ in each step. The perturbation in the n-th step is

$$p^n(x) = \sqrt{\Delta t}\, c(t) \frac{1}{\sqrt{\ell}} \sum_{\nu=1}^{\ell} \xi_\nu^n \sin\left(\frac{2\pi \nu x}{L_0}\right),$$

where the $\xi_\nu^n \in \mathcal{N}(0, 1)$ are normally distributed random numbers with mean zero and standard deviation 1. As in the Euler-Maruyama method [45], the additive noise is scaled with $\sqrt{\Delta t}$. The value ℓ is chosen so that the highest frequency mode has a period $\frac{L_0}{\ell}$ that is not below the vehicle length $1/\rho_{\max}$, i.e., $\ell = \lfloor L_0 \rho_{\max} \rfloor$. In other words, we have white noise exactly until the vehicle scale, which is well-resolved by the numerical scheme. Finally, the noise scale is $c(t) = \frac{1}{100} u_{\max}$ for $t \le 100$, and $c(t) = \frac{1}{1000} u_{\max}$ for $t > 100$. The rationale for this larger initial "thermal noise" is, like in probabilistic optimization techniques, to make it easier for the solutions to escape their initial configuration in case it is only mildly unstable.

Once the solution at t_{final} is found, we first determine the number of shocks. If that number is not equal to 4, we immediately classify the jamiton J_0 as unstable. Otherwise, we check the jamiton speed s by plotting the points $(\rho(x_i, t_{\text{final}}), \rho(x_i, t_{\text{final}}) u(x_i, t_{\text{final}}))$ for $i = 1, \ldots, 10000$ in the fundamental diagram (FD), and calculate s as the least squares best fit slope of these data points (see [56] for the reason why s is the slope in the FD). If $|s - s_0| > 0.5$m/s, we classify J_0 as unstable. Otherwise, we classify J_0 as stable.

This process is now conducted (and run in parallel on a HPC cluster) for 980 different jamitons that are sampled as follows. First we sample 35 values of ρ_S equidistant in the ρ-interval where the SCC is violated. Then, for each ρ_S, we pick 28 values of ρ^+ in $[\rho_S, \rho_M]$, where ρ_M is the upstream density corresponding to the infinite jamiton [56].

The results of this classification are displayed in Fig. 2. Each of the four panels shows the same results, but in four different "phase planes." Each jamiton is uniquely determined by two parameters: (i) the sonic density ρ_S or equivalently the wave speed s; and (ii) the downstream shock density ρ^+, or equivalently, the average density $\bar{\rho}$ across the jamiton, or equivalently, the jamiton length L. Figures 2a, c, and d have the ρ_S on the horizontal axis, and L, ρ^+, and $\bar{\rho}$, respectively, on the vertical axis. Figure 2b displays s vs. $\bar{\rho}$. In each quantity except L, the jamiton region (where the SCC (11) is violated) spans an interval. The dashed brown curve corresponds the zero-length jamiton limit (in which $\rho_S = \rho^+ = \bar{\rho}$), while the solid dark blue curve represents the limit of infinitely long jamitons. Inside that jamiton domain, the 980 investigated jamitons are displayed as colored dots: stable jamitons are light blue; unstable jamitons are red. Note that the void regions visible in Fig. 2a (top left), Fig. 2b (bottom left), and Fig. 2d (bottom right), also possess jamitons that were not simulated due to the sampling strategy of the 980 examples.

The results display intriguingly clear patterns: there appear to be two smooth curves inside the jamiton region that separate the stable from the unstable jamitons.

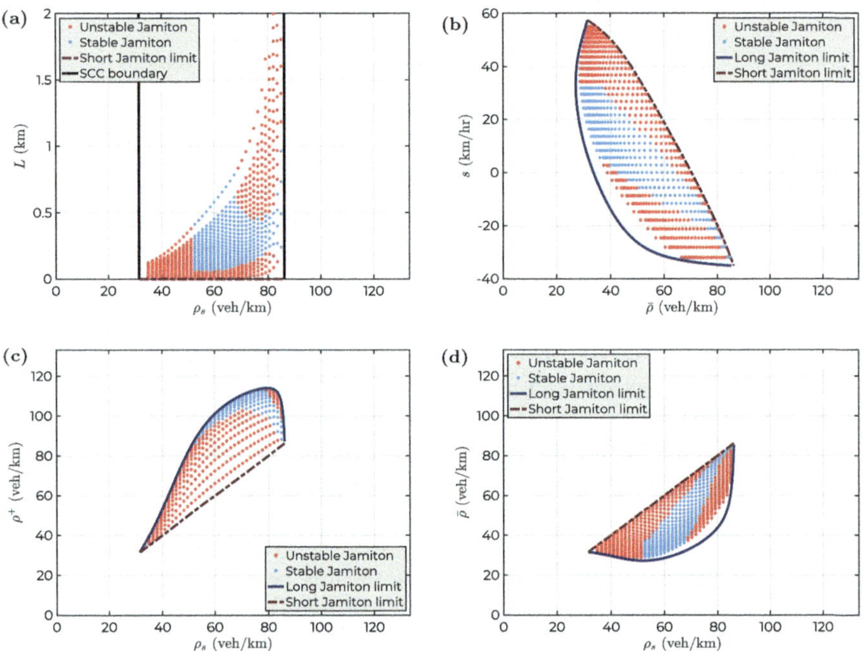

Fig. 2 Classification of 980 jamitons into stable and unstable, displayed in four different phase planes. In each plane, the dashed brown line represents the zero-length jamiton, and the dark blue line is the limit of jamitons with infinite length. The two disconnected red regions correspond to the "splitting" and "merging" instabilities, respectively. (**a**) Stability classification in the phase plane (ρ_S, L). (**b**) Stability classification in the phase plane $(\bar{\rho}, s)$. (**c**) Stability classification in the phase plane (ρ_S, ρ^+). (**d**) Stability classification in the phase plane $(\rho_S, \bar{\rho})$

Specifically, there are two unstable regions separated by a stable region: short jamitons which perturbations cause to coalesce into bigger ones (a "merging" instability); and long jamitons in which the long tail is linearly unstable and sheds growing waves (a "splitting" instability). This last characterization of these two mechanisms is based on observing the time-evolution of the computations, as well as the stability analysis below.

4 Stability Analysis of Jamiton Solutions

We now move towards a mathematical analysis of the dynamic stability of jamitons. For this we switch to the Lagrangian variables introduced in Sect. 2.3. Consider a given jamiton $[\mathrm{v}_0(\sigma, t), u_0(\sigma, t)]^T$ with sonic specific volume v_{s_0}, and Lagrangian length (which is actually the number of vehicles in the jamiton) N_0. We start by writing the (Lagrangian) ARZ model (12) in the frame of reference of this jamiton, which has a propagation speed $-m_0 = h'(\mathrm{v}_{s_0})$. Thus we introduce the variables $\chi = \frac{\sigma + m_0 t}{\tau}$ (the same variable used in Sect. 2.3 to construct the jamitons) and the non-dimensional time $t_* = \frac{t}{\tau}$ (for consistency with the scaling used for χ). Because of that last choice, any instability growth rate computed with these variables needs to be scaled by τ to recover physical units.

In the coordinates defined above, Eqs. (12) become

$$\begin{aligned} \mathrm{v}_{t_*} + (m_0 \mathrm{v} - u)_\chi &= 0 , \\ (u + h(\mathrm{v}))_{t_*} + m_0 \, (u + h(\mathrm{v}))_\chi &= U(\mathrm{v}) - u . \end{aligned} \tag{18}$$

This system is in conservative form, with conserved quantities v and $q = u + h(\mathrm{v})$. The characteristic speeds of (18) are

$$\lambda_1 = m_0 + h'(\mathrm{v}) \quad \text{and} \quad \lambda_2 = m_0 . \tag{19}$$

The Rankine-Hugoniot shock jump conditions associated with (18) are

$$\begin{aligned} (-m_0 + \tilde{m}) \, [\mathrm{v}] + [u] &= 0 , \\ [u] + [h(\mathrm{v})] &= 0 , \end{aligned} \tag{20}$$

where $\tilde{m}$ is the shock speed in the χ–t_* frame. Contacts require $\tilde{m} = m_0$ and $[u] = 0$.

4.1 Perturbation System for Single-Jamiton Waves

We now formulate a linear perturbation system of (18). There are two fundamental differences to the linear perturbation analysis for uniform flow presented in

Sect. 2.2. First, because the jamiton profile is non-constant, we obtain a variable-coefficient linear system. Second, because the jamiton contains a shock, we must introduce a perturbation to the shock's position as an additional variable (a variable not needed for perturbations of smooth solutions). As we will see below in more detail, both aspects render this analysis significantly more complicated than the one in Sect. 2.2.

Here we consider the stability of periodic jamiton profiles with one shock per period, under periodic perturbations. Note that this setup excludes the possibility of jamitons merging by means of adjacent shocks approaching each other. Hence, we only study the "splitting instability" for long jamitons, not the "merging instability" for short jamitons (see Sect. 3.2).

Consider a periodic jamiton profile $[\mathrm{v}_0(\sigma, t), u_0(\sigma, t)]^T$ of length N_0 between shocks, and write it as $[\mathrm{v}_0(\chi), u_0(\chi)]^T$—a solution of (18) on $[0, N_0]$ with the shock placed at 0. Now write $\mathrm{v}(\chi, t_*) = \mathrm{v}_0(\chi) + \delta\mathrm{v}(\chi, t_*)$ and $u(\chi, t_*) = u_0(\chi) + \delta u(\chi, t_*)$, where $\delta\mathrm{v}$ and δu are infinitesimal perturbations. Substituting into (18) yields the linear system for $\delta\mathrm{v}$ and δu:

$$\begin{aligned} \delta\mathrm{v}_{t_*} + (m_0\delta\mathrm{v} - \delta u)_\chi &= 0\,, \\ \left(\delta u + h'(\mathrm{v}_0)\delta\mathrm{v}\right)_{t_*} + m_0\left(\delta u + h'(\mathrm{v}_0)\delta\mathrm{v}\right)_\chi &= U'(\mathrm{v}_0)\delta\mathrm{v} - \delta u\,. \end{aligned} \tag{21}$$

We also need to track the infinitesimal perturbation of the shock position $\chi = \mu(t_*)$. We do so by implementing the Rankine-Hugoniot conditions (20) in a way consistent with solving (18) on $[0, N_0]$ with periodic boundary conditions. This then generates boundary conditions for (21). The first equation in (20) yields

$$(\dot{\mu} - m_0)\Big([\mathrm{v}_0] + [\delta\mathrm{v}] + \mu[\mathrm{v}_{0\chi}]\Big) + [u_0] + [\delta u] + \mu[u_{0\chi}] = 0\,.$$

Expanding this equation, ignoring terms beyond $O(\mu)$, and using that $[u_0] - m_0[\mathrm{v}_0] = 0$ and $u_{0\chi} - m_0\mathrm{v}_{0\chi} = 0$, we obtain

$$\dot{\mu}[\mathrm{v}_0] - m_0[\delta\mathrm{v}] + [\delta u] = 0\,.$$

The second equation in (20) becomes

$$[u_0] + [\delta u] + \mu[u_{0\chi}] + [\mathrm{v}_0] + [h'(\mathrm{v}_0)\delta\mathrm{v}] + \mu[h(\mathrm{v}_0)_\chi] = 0\,.$$

Again, ignoring terms beyond $O(\mu)$ and using that $[u_0] + [h(\mathrm{v}_0)] = 0$, we get

$$[\delta u] + [h'(\mathrm{v}_0)\delta\mathrm{v}] + \mu[u_{0\chi} + h'(\mathrm{v}_0)_\chi] = 0\,.$$

In this setup the bracket notation denotes $[\zeta] = \zeta(0^+) - \zeta(N_0^-)$. Therefore, we have derived the following variable-coefficient linear model for $\delta\mathrm{v}$ and δu on $[0, N_0]$, with boundary conditions that involve the shock position perturbation μ:

$$
\begin{aligned}
\delta \mathrm{v}_{t_*} + (m_0 \delta \mathrm{v} - \delta u)_\chi &= 0 \,, \\
\big(\delta u + h'(\mathrm{v}_0)\delta \mathrm{v}\big)_{t_*} + m_0\big(\delta u + h'(\mathrm{v}_0)\delta \mathrm{v}\big)_\chi &= U'(\mathrm{v}_0)\delta \mathrm{v} - \delta u \,, \qquad (22)\\
\text{with boundary condition:} \quad \big[\delta u + h'(\mathrm{v}_0)\delta \mathrm{v}\big] &= -\mu \big[u_{0\chi} + h'(\mathrm{v}_0)_\chi\big] \,, \\
\text{where } \mu \text{ satisfies the ODE:} \qquad \dot{\mu} &= \frac{m_0[\delta \mathrm{v}] - [\delta u]}{[\mathrm{v}_0]} \,. \qquad (23)
\end{aligned}
$$

We conduct two further simplifications to the model. First, we transform it to characteristic form by writing it in terms of the Riemann variables δu and $\delta q = \delta u + h'(\mathrm{v}_0)\delta \mathrm{v}$. Second, we replace the shock perturbation variable μ by a Robin b.c. for the PDE, as follows. Differentiating the boundary conditions $[\delta q] = -\mu[u_{0\chi} + h'(\mathrm{v}_0)_\chi]$ with respect to time yields

$$
\begin{aligned}
\frac{d}{dt_*}[\delta q] &= -\big[u_{0\chi} + h'(\mathrm{v}_0)_\chi\big]\,\dot{\mu} \\
&= \frac{-\big[u_{0\chi} + h'(\mathrm{v}_0)_\chi\big]}{[\mathrm{v}_0]} \left(\left[\frac{m_0}{h'(\mathrm{v}_0)}\delta q\right] - \left[\left(1 + \frac{m_0}{h'(\mathrm{v}_0)}\right)\delta u\right]\right).
\end{aligned}
$$

Using the fact that $\delta q_{t_*} = -m_0 \delta q_\chi + \left(\frac{-h'(\mathrm{v}_0) - U'(\mathrm{v}_0)}{h'(\mathrm{v}_0)}\right)\delta u + \left(\frac{U'(\mathrm{v}_0)}{h'(\mathrm{v}_0)}\right)\delta q$, we obtain Robin boundary conditions for the PDE. Altogether, we obtain the following system

$$
\begin{aligned}
\delta u_{t_*} + \big(m_0 + h'(\mathrm{v}_0)\big)\delta u_\chi &= \left(\frac{m_0 h'(\mathrm{v}_0)_\chi - h'(\mathrm{v}_0) - U'(\mathrm{v}_0)}{h'(\mathrm{v}_0)}\right)\delta u + \left(\frac{U'(\mathrm{v}_0) - m_0 h'(\mathrm{v}_0)_\chi}{h'(\mathrm{v}_0)}\right)\delta q \,, \\
\delta q_{t_*} + m_0 \delta q_\chi &= \left(\frac{-h'(\mathrm{v}_0) - U'(\mathrm{v}_0)}{h'(\mathrm{v}_0)}\right)\delta u + \left(\frac{U'(\mathrm{v}_0)}{h'(\mathrm{v}_0)}\right)\delta q \,,
\end{aligned}
\qquad (24)
$$

with boundary condition

$$
\delta q_\chi(0) + k_{\mathrm{L}}\delta q(0) = \delta q_\chi(N_0) + k_{\mathrm{R}}\delta q(N_0) + c_{\mathrm{L}}\delta u(0) + c_{\mathrm{R}}\delta u(N_0) \,. \qquad (25)
$$

The coefficients are computable from the jamiton functions as

$$
k_{\mathrm{L}} = K(0) \,, \quad k_{\mathrm{R}} = K(N_0) \,, \quad c_{\mathrm{L}} = -C(0) \,, \quad \text{and} \quad c_{\mathrm{R}} = C(N_0) \,,
$$

where

$$
\begin{aligned}
K(\chi) &= \frac{1}{m_0} - \frac{[u_{0\chi} + h'(\mathrm{v}_0)_\chi]}{[\mathrm{v}_0]} \frac{1}{h'(\mathrm{v}_0(\chi))} - \frac{h'(\mathrm{v}_0(\chi)) + U'(\mathrm{v}_0(\chi))}{m_0 h'(\mathrm{v}_0(\chi))} \,, \\
C(\chi) &= \frac{[u_{0\chi} + h'(\mathrm{v}_0)_\chi]}{[\mathrm{v}_0]} \left(\frac{1}{m_0} + \frac{1}{h'(\mathrm{v}_0(\chi))}\right) + \frac{h'(\mathrm{v}_0(\chi)) + U'(\mathrm{v}_0(\chi))}{m_0 h'(\mathrm{v}_0(\chi))} \,.
\end{aligned}
$$

4.2 Qualitative Characterization of the Jamiton Perturbation System

We now adopt a short notation for the jamiton perturbation system (24), with b.c. (25), by writing (u, q) and (x, t) in place of $(\delta u, \delta q)$ and (χ, t_*), and introducing coefficient functions to obtain

$$\begin{aligned} u_t + b_1(x)u_x &= a_{11}(x)u + a_{12}(x)q \ , \\ q_t + b_2 \quad q_x &= a_{21}(x)u + a_{22}(x)q \ , \end{aligned} \tag{26}$$

with b.c. $(q_x + k_{\rm L}q)(0) = (q_x + k_{\rm R}q)(N_0) + c_{\rm L}u(0) + c_{\rm R}u(N_0)$. The characteristic speed $b_2 > 0$ is constant and positive. In turn, $b_1(x)$ vanishes at the sonic point $x_{\rm S}$, and is negative (positive) for $x < x_{\rm S}$ $(x > x_{\rm S})$. Hence, the only in-going characteristic is at $x = 0$, for q (consistent with a single b.c.). The function $a_{11}(x)$ crosses from negative to positive at $x_{\rm S}$ as well, and it is always negative for $x < x_{\rm S}$; it may or may not cross back to negative for some $x > x_{\rm S}$. Finally, $a_{22}(x) < 0$ everywhere. Figures 3 and 4 display the functions and characteristic curves, respectively, for an example jamiton.

Qualitatively, the solutions of (26) behave as follows. Being an advection-reaction system, its solutions are generally wave-like in nature. Waves enter the q-field at $x = 0$ and are transported with the q-field to the right with constant speed b_2, while being dampened by the a_{22}-term and modified (via the u-field) through the a_{21}-term. Likewise, the q-field constantly feeds into the u-field via the a_{12}-term. Moreover, for $x < x_{\rm S}$, the u-field is transported towards $x = 0$ and dampened by a_{11}; while for $x > x_{\rm S}$, the u-field is transported towards $x = N_0$ and amplified/dampened by a_{11}. Finally, the outgoing characteristics at $x = 0$ (u) and $x = N_0$ (u and q) combine via (25) and feed back into q at $x = 0$.

Our goal is now to (a) characterize the dynamic stability of the given jamiton by means of the behavior of the solutions of its associated perturbation system (26) (incl. b.c.), and (b) use this insight to explain and understand the computational results of the fully nonlinear ARZ model (4) presented in Sect. 3. To that end, we

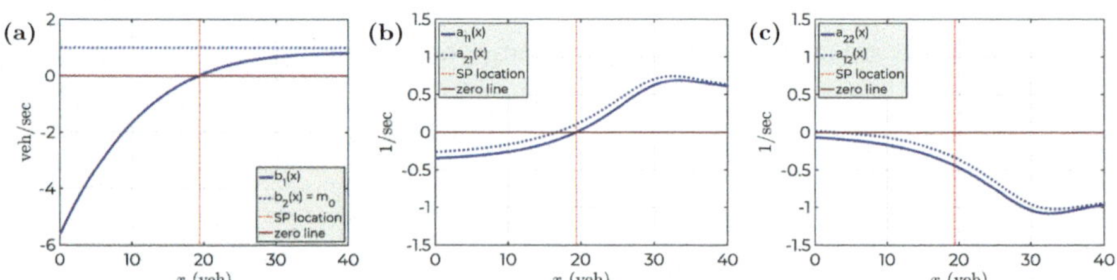

Fig. 3 Coefficient functions for (26) and a jamiton with $v_{\rm S} = 12.5$ m/veh and $v^+ = 8.9$ m/veh. This jamiton has a length of 561m and contains 40 vehicles. Note that here we revert to physical units (vehicles) for the horizontal axis. (**a**) $b_1(x)$ and b_2. (**b**) $a_{11}(x)$ and $a_{21}(x)$. (**c**) $a_{22}(x)$ and $a_{12}(x)$

start by establishing that there are (at least) two distinct notions of (in)stability that must be considered here.

First, *asymptotic stability* under infinitesimal perturbations (studied in Sect. 4.4). This is captured by the $t \to \infty$ behavior of linear model (26): if for any i.c. $[u(x,0), q(x,0)]^T$ the solution decays exponentially as $t \to \infty$, then this notion of stability is met. *Strong linear instability* occurs when there is a positive feedback mechanism that produces an exponential growth of an initial perturbation in time, eventually driving the full model (4) out of the linear regime, no matter how small the initial (non-zero) perturbation is. At the borderline between these two behaviors, the solutions to the linear system may remain bounded for all time, or grow/decay at a sub-exponential rate.

The second notion of stability is given by the *maximum transient growth* criteria (studied in Sect. 4.5). Because (26) is non-normal, even if asymptotic stability applies, an initially small perturbation may be amplified significantly at transient times, before eventually dying off as $t \to \infty$. However, if that amplified perturbation becomes sufficiently large, nonlinear effects will take over in the full ARZ model (4). In this scenario, how far the system ends up from equilibrium depends both on the transient growth factor (see below) and the magnitude of the perturbations.

4.3 Fundamental Challenges Caused by the Sonic Point

In the same way as the original inhomogenous ARZ model may look misleadingly innocuous ("just a hyperbolic system with a relaxation term"), yet develops extremely complex dynamics if the SCC is violated, the jamiton perturbation system (24) may look innocent as well—and also that impression would be false. The fact that the characteristic speed b_1 transitions from negative to positive at x_S (a direct consequence of this being a sonic point), causes fundamental structural challenges.

It may seem rather natural to attempt to study (24) by expanding its solutions using eigenmodes, and seek solutions to the eigenvalue problem

$$\begin{cases} \lambda u &= -b_1(x)u_x + a_{11}(x)u + a_{12}(x)q \ , \\ \lambda q &= -b_2 q_x + a_{21}(x)u + a_{22}(x)q \ . \end{cases} \tag{27}$$

However, the right-hand side operator here is non-normal; and it is well known that for non-normal operators, spectral calculations can be extremely unreliable [14, 60, 61].

Furthermore, the presence of the sonic point makes the situation substantially worse, even if one were to have access to "exact" computations. To illustrate the issue consider the simple model problem

$$u_t + (xu)_x = \tfrac{3}{4}u \ , \quad -1 < x < 1 \ . \tag{28}$$

The exact solution of (28) is easily obtained using characteristics: $u = u_0(x\,e^{-t})e^{-\frac{1}{4}t}$, where u_0 is the initial data. This *clearly* is a stable situation by any "physically reasonable" definition. On the other hand, if we look for eigenfunctions by separating $u = \phi(x)e^{\lambda t}$, we find that: $\phi = |x|^{\alpha}$, with $\alpha = -(\lambda + \frac{1}{4})$ and *any* λ with $\mathrm{Re}(\lambda) < \frac{1}{4}$, is an acceptable square-integrable eigenfunction. Even worse: every eigenvalue has infinite multiplicity (apply $\frac{\mathrm{d}^n}{\mathrm{d}\alpha^n}$ to the eigenvalue equation with the solutions above).

Thus from a naive eigenvalue calculation one would conclude that an exponential instability occurs! But here, with an exact solution, the situation is clear: the presence of a sonic point allows the existence of solutions that are not smooth. Then stability and growth/decay rates depend on the smoothness restrictions imposed. While L^2 yields instability, L^{∞} or H^1 yield stability, but with different bounds on the decay rates. Thus, in a numerical computation one would have to worry about what restriction (if any) the computation enforces as the resolution increases.

Because of these issues we refrain from using the approach in (27), and instead characterize (in)stability via alternative ways that do not use eigenmode expansions.

4.4 Quantitative Results: Asymptotic Stability

The $t \to \infty$ behavior of the solutions of the jamiton perturbation system (26) (incl. b.c.) depends on a delicate balance of growth vs. decay effects. And because those are governed by the functions $a_{ij}(x)$, $b_i(x)$, and the b.c. constants, we do not attempt a fully analytical characterization here. Instead, we formulate a sequence of approximations to the solutions of (26) and analyze their behavior. Specifically, we formulate the following approximation scheme.

We discretize the spatial domain into a regular grid $\{x_0, \ldots, x_m\} = \{0, h, 2h, \ldots, N_0 - h, N_0\}$ and conduct time steps of size $\Delta t = h/b_2$, see Fig. 4. We denote the grid approximations $U_j^n \approx u(jh, n\Delta t)$ and $Q_j^n \approx q(jh, n\Delta t)$, and denote the full state vector at time $n\Delta t$ by $\mathbf{Y}^n = [\mathbf{U}^n, \mathbf{Q}^n]^T$, where $\mathbf{U}^n = [U_1^n, \ldots, U_m^n]^T$ and $\mathbf{Q}^n = [Q_1^n, \ldots, Q_m^n]^T$. An update matrix for the

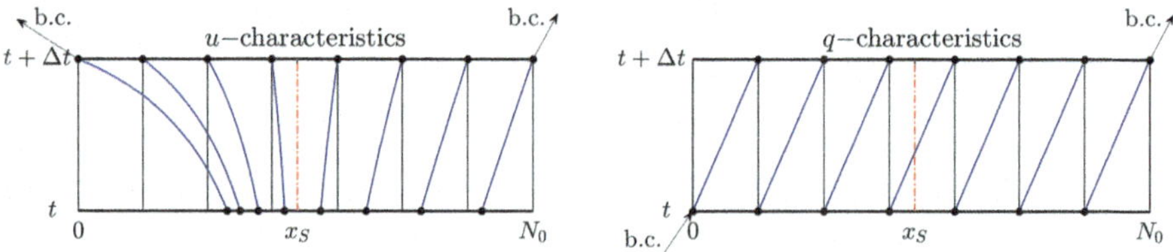

Fig. 4 Illustration of the discretization used to approximate (26), as described in Sect. 4.4. The left (right) graphic shows the characteristic curves corresponding to the u (q) variable. The u-characteristics expand away from the sonic point towards the domain boundaries (where the shock is). The scheme's time step is selected so that the q-characteristics advance by h per time step

transport part of (26) (incl. b.c., but neglecting the a_{ij}-terms) is obtained via tracking characteristics: for each grid point $x_j = jh$, determine the associated foot point $\mathring{x}_j$ as the solution of the ODE $\dot{x}(s) = -b_1(x(s))$ with $x(0) = x_j$, evaluated at $s = \Delta t$. Then, $U_j^{n+1} = \Psi_{\mathbf{U}^n}(\mathring{x}_j)$, where $\Psi_{\mathbf{U}^n}(x)$ is the piecewise-linear interpolant based on the grid data $\mathbf{U}^n$. Due to the clever choice of time step, the q-update can be solved exactly via $Q_j^{n+1} = Q_{j-1}^n$ for all $j \geq 1$. The b.c. are used to update $Q_0^{n+1} = \frac{1}{k_L - h^{-1}}((k_R - h^{-1})Q_{m-1}^n + c_L\Psi_{\mathbf{U}^n}(\mathring{x}_0) + c_R\Psi_{\mathbf{U}^n}(\mathring{x}_m))$. We denote this update matrix M_1.

A second matrix for the growth/decay part (i.e., neglecting the advection terms) is formulated as follows: $[U_j^{n+1}, Q_j^{n+1}]^T = \exp(\Delta t A(x_j)) \cdot [U_j^n, Q_j^n]^T$, where $A(x)$ is the 2×2 matrix formed by the $a_{ij}(x)$ values. We denote the resulting update matrix M_2.

One step of the numerical scheme, $\mathbf{Y}^{n+1} = M \cdot \mathbf{Y}^n$, is given by the update matrix $M = M_2 \cdot M_1$. This first-order method is carefully designed to not incur any slow drifts. Because the scheme is linear with time-independent coefficients, the $t \to \infty$ behavior of the solutions is fully characterized by its one-step update matrix M, specifically by its spectral radius $\rho(M)$: asymptotic stability (of the approximation) is given exactly if $\rho(M) < 1$. Once M is set up, this stability condition can be checked via Matlab's numerical linear algebra routines, resulting in a systematic classification of jamitons into asymptotically stable vs. unstable.

A caveat in this approach is that for any choice of grid size h, we check the asymptotic stability of an *approximation* to (26). However, because we have a convergent sequence of approximations, we approach the true answer for (26) as $h \to 0$. Moreover, for any $h > 0$, the approximation slightly overestimates stability due to the scheme's numerical diffusion (which vanishes as $h \to 0$), resulting in a too small but growing (as $h \to 0$) unstable jamiton region.

Figure 5 displays the results. It shows the classification of the same jamitons as in Fig. 2 into asymptotically stable and unstable using the asymptotic stability criterion: $\rho(M) < 1$ (unstable: $\rho(M) > 1$), where for each jamiton, M is the one-step update matrix that comes from a discretization with 8000 grid points. Comparing those results to the nonlinear system results in Fig. 2, we indeed see that (i) only the splitting instability (long jamitons) can be captured; and (ii) the unstable region is underestimated. This last aspect is likely also affected by the fact that asymptotic stability does not account for transient growth effects; which we consider next.

4.5 *Quantitative Results: Transient Growth*

Even if the system (26) is asymptotically stable, small perturbations may be amplified significantly at transient times. Via asymptotic arguments we can argue that the dominant wave amplitude growth mechanism is the growth of the u-field

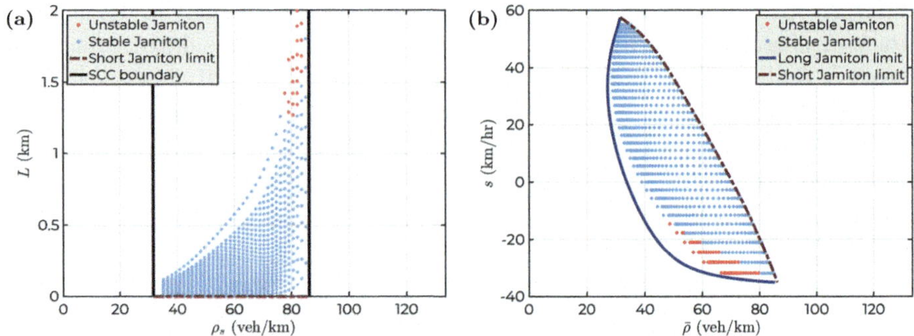

Fig. 5 Classification of 980 jamitons into asymptotically stable vs. unstable, where asymptotic stability is given by $\rho(M) < 1$ (and instability by $\rho(M) > 1$). Here, for each jamiton, M is the one-step update matrix that comes from a discretization with 8000 grid points. Note that the criterion used here can only detect "splitting" instabilities. (**a**) Classification of asymptotic stability in the phase plane (ρ_S, L). (**b**) Classification of asymptotic stability in the phase plane $(\bar{\rho}, s)$

as it travels between the sonic point x_S and the right domain boundary N_0. The argument (which can be made rigorous via a WKB expansion [6]) is as follows.

Consider high frequency solutions of (26), i.e., solutions that are rapidly varying in space and time. In this situation the behavior is dominated by the left-hand side, and we can see that such solutions generally consist of a superposition of two waves: the "u-wave," dominated by the excitation in u, and the "q-wave," dominated by the excitation in q. Consider first the u-wave. Then, because $u \gg q$, we can simplify the equations to obtain

$$u_t + b_1(x)u_x \approx a_{11}(x)u \ ,$$
$$q_t + b_2 \quad q_x \approx a_{21}(x)u \ .$$

From this we can see that q is "slaved" to u (since the homogeneous part of the solution to the second equation should be considered as belonging to the q-wave). A similar argument applies to the q-wave; however, the u-wave will dominate because $a_{11} > 0$ to the right of x_S, while $a_{22} < 0$.

Hence, neglecting the q-wave (and its influence on u) we obtain that u evolves (approximately) according to the characteristic equations $\frac{\mathrm{d}x}{\mathrm{d}t} = b_1(x)$ and $\frac{\mathrm{d}u}{\mathrm{d}t} = a_{11}(x)u$. The speed b_1 vanishes at x_S, but so does the growth rate a_{11}, resulting in an overall finite net growth. By the chain rule, the characteristic equations lead to the ODE $\frac{\mathrm{d}u}{\mathrm{d}x} = \frac{a_{11}(x)}{b_1(x)}u$, with normalized i.c. $u(x_S) = 1$, to estimate the transient amplification factor F. Solving the ODE yields

$$F = \exp\left(\int_{x_S}^{N_0} \frac{a_{11}(x)}{b_1(x)}\,\mathrm{d}x\right) \ . \tag{29}$$

This quantity can be computed via quadrature, using L'Hôpital's rule at/near x_S. However, note that the arguments above do not apply across the sonic point, even though the integrand is not singular, because the parameterization of the characteristics by x (i.e., $\frac{\mathrm{d}t}{\mathrm{d}x} = \frac{1}{b_1(x)}$) implicit in the calculation above breaks down there.

An important fact is that the quantity F can be computed without solving the jamiton ODE. This is achieved by parameterizing the jamiton in terms of v_S and the left shock state $\mathrm{v}_{N_0} = \mathrm{v}(N_0)$. Then, because a_{11} and b_1 are functions of x only via the jamiton $\mathrm{v}(x)$, one can apply a change of variables to replace x-integration by v-integration. The Jacobian for the transformation follows from the jamiton ODE (17). This yields the formula

$$F = \exp\left(\int_{\mathrm{v}_S}^{\mathrm{v}_{N_0}} \frac{m_0 h''(\mathrm{v})}{h'(\mathrm{v})(h'(\mathrm{v}) + m_0)} - \frac{m_0(h'(\mathrm{v}) + U'(\mathrm{v}))}{h'(\mathrm{v})(U(\mathrm{v}) - m_0\mathrm{v} - s_0)} \,\mathrm{d}\mathrm{v}\right) .$$

Figure 6 shows the stability classification via this criterion for the same jamitons studied in Fig. 2. As in Fig. 5, we do not capture merging instabilities. For the splitting instability, we consider two thresholds for the amplification factor: $F_1 = 10^5$ and $F_2 = 10^{15}$. Classifying jamitons below the 10^5 amplification factor as stable is consistent with the magnitude of noise in the nonlinear computation (Sect. 3.2), which was roughly 10^{-5}. The results show that the stability boundary in Fig. 2 is not reproduced perfectly, but reasonably well. An interesting advantage of this measure of "instability" is that it is not just a yes/no criterion, but rather provides a measure of the "badness" of the instability. One key missing piece in this criterion is that it does not characterize the "pumping" mechanism of perturbations from q into u at/near the sonic point. Hence, we do not know how large the perturbation magnitude really is near x_S.

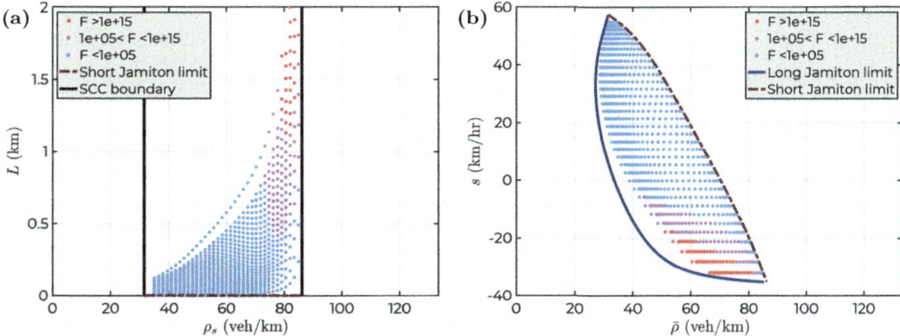

Fig. 6 Classification of 980 jamitons according to the transient growth factor (29). Three levels of F are displayed, with the thresholds at $F_1 = 10^5$ and $F_2 = 10^{15}$ to yield: stable if $F < F_1$, moderately unstable if $F_1 < F < F_2$, and unstable if $F_2 < F$. (**a**) Classification of jamitons according to F in the phase plane (ρ_S, L). (**b**) Classification of jamitons according to F in the phase plane $(\bar{\rho}, s)$

5 Discussion and Outlook

The study presented in this paper highlights important structural properties of hyperbolic conservation law systems with relaxation terms, in the regime when the sub-characteristic condition (SCC) is violated. Such PDE are of importance in the macroscopic modeling of vehicular traffic flow (the main focus here), but also for other applications, such as roll waves in open channels [49] and circular hydraulic jumps [30]. Furthermore, many of the issues are similar to those that appear in the context of the ZND theory for the stability of Chapman-Jouguet (CJ) detonations [18]. In fact, jamitons are mathematical analogs of detonation waves [19]. While for detonation waves the notion of an SCC does not seem to apply, CJ detonations do have a sonic point, which renders their stability analysis [8, 58] difficult. It is our hope that the relative simplicity of systems such as the ARZ model will provide a route to advance in this challenging topic.

This work provides a pathway to understanding important stability questions for the inhomogeneous ARZ model (3). In the regime of violated SCC, this model can reproduce the practically relevant [57] phenomena of phantom traffic jams and stop-and-go traffic waves, while preserving the advantages of a macroscopic description (see Sect. 1). The dynamic stability of jamitons determines which of the many theoretically possible jamiton solutions of the model can/will be selected by the equations' dynamics. The study in Sect. 3 reveals that short jamitons tend to merge, and long jamitons tend to split, resulting in a middle range of stable jamiton wave lengths. A remarkable aspect about this dynamic selection via (in)stability is that it selects a length scale (range), even though there is no length scale that is explicitly inserted into the model.

The perturbation analysis of jamiton solutions presented here leads to a variable-coefficient linear advection-reaction system whose solutions characterize jamiton stability. As shown in Sect. 4, this system exhibits extremely complex dynamics that may not be suspected at first glance, given its simple fundamental structure. A key reason for those complex dynamics is the zero-transition of one characteristic field, which corresponds to the sonic point in the nonlinear jamiton. While a complete analysis of the behavior of the solutions to the perturbation system remains to be conducted in future work (including a full WKB analysis [6]), the qualitative characterization presented herein reveals that there are two key mechanisms for instability that must be considered: first, asymptotic stability that captures the net amplification or decay of infinitesimal perturbation that traverse through periodic jamiton patterns; and second, the transient growth of small perturbations as they travel from near the sonic point down the jamiton profile until they eventually hit the next shock. The quantitative study in Sect. 4 reveals that for some jamitons, such transient amplifications may yield noise amplification by many orders of magnitude, which for many practical situations will definitely push the solutions into the fully nonlinear regime.

Based on those stability concepts, two criteria have been developed that are directly verifiable in terms of the model functions rather than requiring nonlinear hyperbolic system simulations. Asymptotic stability reduces to finding the spectral radius of a sparse matrix, which in itself is a nontrivial problem as well, but it is an established standard task in numerical linear algebra. For the transient growth, a proxy criterion has been devised that boils down to a straightforward quadrature of two model functions. When compared with the "brute force" nonlinear stability results (Sect. 3), those two criteria capture the key qualitative essence of the stability boundary for long jamitons; but to reproduce the precise shape there is still room for improvement via more refined stability criteria.

Mathematically, understanding the solution behavior of relaxation system in which the SCC is violated is a crucial challenge [29, 41, 44], and this work provides some insight. In addition, the jamiton perturbation system (24) is full of challenging structure (see Sect. 4.3), and this paper provides criteria to characterize its stability properties.

For the key application of traffic flow, the understanding of which jamiton solutions are dynamically stable is a critical step towards determining which models reproduce real-world phenomena best. Moreover, the non-normal structure of the system in (24), leading to the transient growth behavior it exhibits (Sect. 4.5), has interesting connections to the task of stabilizing traffic flow with a single autonomous vehicle [10].

Finally, an obvious extension is to tackle the merging instability as well, and we plan to do so in future work. At least in principle, the methodology of this current work can be extended to include the merging instabilities by allowing multiple shock perturbation.

Acknowledgments The authors would like to acknowledge the support by the National Science Foundation. R. R. Rosales and B. Seibold were supported through grants DMS–1719637 and DMS–1719640, respectively. B. Seibold and R. Ramadan were also supported by DOE. This material is based upon work supported by the U.S. Department of Energy's Office of Energy Efficiency and Renewable Energy (EERE) award number CID DE-EE0008872. The views expressed herein do not necessarily represent the views of the U.S. Department of Energy or the United States Government. Computations were carried out on Temple University's HPC resources and thus were supported in part by the National Science Foundation through major research instrumentation grant number 1625061.

References

1. T. Alperovich, A. Sopasakis, Modeling highway traffic with stochastic dynamics. J. Stat. Phys **133**, 1083–1105 (2008)
2. S. Amin et al., Mobile century — Using GPS mobile phones as traffic sensors: A field experiment, in *15th World Congress on Intelligent Transportation Systems*, New York, Nov. 2008
3. A. Aw, M. Rascle, Resurrection of second order models of traffic flow. SIAM J. Appl. Math. **60**, 916–944 (2000)
4. A. Aw, A. Klar, T. Materne, M. Rascle, Derivation of continuum traffic flow models from microscopic follow-the-leader models. SIAM J. Appl. Math. **63**(1), 259–278 (2002)

5. M. Bando, K. Hesebem, A. Nakayama, A. Shibata, Y. Sugiyama, Dynamical model of traffic congestion and numerical simulation. Phys. Rev. E **51**(2), 1035–1042 (1995)
6. C. Bender, S. Orszag, *Advanced Mathematical Methods for Scientists and Engineers* (McGraw-Hill, New York, 1978)
7. G.Q. Chen, C.D. Levermore, T.P. Liu, Hyperbolic conservation laws with stiff relaxation terms and entropy. Comm. Pure Appl. Math. **47**, 787–830 (1994)
8. P. Clavin, B. Benet, Decay of plane detonation waves to the self-propagating Chapman—Jouget regime. J. Fluid Mech. **845**, 170–202 (2018)
9. R.M. Colombo, On a 2×2 hyperbolic traffic flow model. Math. Comput. Modell. **35**, 683–688 (2002)
10. S. Cui, B. Seibold, R.E. Stern, D.B. Work, Stabilizing traffic flow via a single autonomous vehicle: Possibilities and limitations, in *Proceedings of the 2017 IEEE Intelligent Vehicles Symposium*, Redondo Beach, 2017
11. C.F. Daganzo, The cell transmission model: A dynamic representation of highway traffic consistent with the hydrodynamic theory. Transp. Res. B **28**, 269–287 (1994)
12. C.F. Daganzo, Requiem for second-order fluid approximations of traffic flow. Transp. Res. B **29**, 277–286 (1995)
13. P.G. Drazin, W.H. Reid, *Hydrodynamic Stability* (Cambridge University Pres, 1981)
14. M. Embree, L.N. Trefethen, Generalizing eigenvalue theorems to pseudospectra theorems. SIAM J. Sci. Comput. **23**(2), 583–590 (2001)
15. L.C. Evans, *Partial Differential Equations*, volume 19 of *Graduate Studies in Mathematics* (American Mathematical Society, 1998)
16. S. Fan, M. Herty, B. Seibold, Comparative model accuracy of a data-fitted generalized Aw-Rascle-Zhang model. Netw. Heterog. Media **9**(2), 239–268 (2014)
17. S. Fan, B. Seibold, Data-fitted first-order traffic models and their second-order generalizations: Comparison by trajectory and sensor data. Transp. Res. Rec. **2391**, 32–43 (2013)
18. W. Fickett, W.C. Davis, *Detonation* (Univ. of California Press, Berkeley, CA, 1979)
19. M.R. Flynn, A.R. Kasimov, J.-C. Nave, R.R. Rosales, B. Seibold, Self-sustained nonlinear waves in traffic flow. Phys. Rev. E **79**(5), 056113 (2009)
20. H. Greenberg, An analysis of traffic flow. Oper. Res. **7**, 79–85 (1959)
21. J.M. Greenberg, Extension and amplification of the Aw-Rascle model. SIAM J. Appl. Math. **63**, 729–744 (2001)
22. B.D. Greenshields, A study of traffic capacity. Proc. Highw. Res. Rec. **14**, 448–477 (1935)
23. A. Harten, High resolution schemes for hyperbolic conservation laws. J. Comput. Phys. **49**, 357–393 (1983)
24. D. Helbing, Traffic and related self-driven many-particle systems. Rev. Modern Phys. **73**, 1067–1141 (2001)
25. D. Helbing, A.F. Johansson, On the controversy around Daganzo's requiem for and Aw-Rascle's resurrection of second-order traffic flow models. Eur. Phys. J. B **69**(4), 549–562 (2009)
26. R. Herman, I. Prigogine, *Kinetic Theory of Vehicular Traffic* (Elsevier, New York, 1971)
27. J.-C. Herrera, D. Work, X. Ban, R. Herring, Q. Jacobson, A. Bayen, Evaluation of traffic data obtained via GPS-enabled mobile phones: The mobile century field experiment. Transp. Res. B **18**, 568–583 (2010)
28. R. Illner, A. Klar, T. Materne, Vlasov-Fokker-Planck models for multilane traffic flow. Commun. Math. Sci. **1**(1), 1–12 (2003)
29. S. Jin, M.A. Katsoulakis, Hyperbolic systems with supercharacteristic relaxations and roll waves. SIAM J. Appl. Math. **61**, 273–292 (2000)
30. A.R. Kasimov, A stationary circular hydraulic jump, the limits of its existence and its gasdynamic analogue. J. Fluid Mech. **601**, 189–198 (2008)
31. B.S. Kerner, Experimental features of the emergence of moving jams in free traffic flow. J. Phys. A **33**, 221–228 (2000)
32. B.S. Kerner, P. Konhäuser, Cluster effect in initially homogeneous traffic flow. Phys. Rev. E **48**, R2335–R2338 (1993)

33. B.S. Kerner, P. Konhäuser, Structure and parameters of clusters in traffic flow. Phys. Rev. E **50**, 54–83 (1994)
34. T.S. Komatsu, S. Sasa, Kink soliton characterizing traffic congestion. Phys. Rev. E **52**, 5574–5582 (1995)
35. D.A. Kurtze, D.C. Hong, Traffic jams, granular flow, and soliton selection. Phys. Rev. E **52**, 218–221 (1995)
36. P.D. Lax, *Hyperbolic Systems of Conservation Laws and the Mathematical Theory of Shock Waves*, vol. 11 (SIAM, 1973)
37. J.-P. Lebacque, Les modeles macroscopiques du traffic. Annales des Ponts. **67**, 24–45 (1993)
38. J.-P. Lebacque, S. Mammar, H. Haj-Salem, Generic second order traffic flow modelling, in *Transportation and Traffic Theory*, ed. by R.E. Allsop, M.G.H. Bell, B.G. Heydecker. Proc. of the 17th ISTTT (Elsevier, 2007), pp. 755–776
39. R.J. LeVeque, *Numerical Methods for Conservation Laws*, 2nd edn. (Birkhäuser, 1992)
40. R.J. LeVeque, *Finite Volume Methods for Hyperbolic Problems*, 1st edn. (Cambridge University Press, 2002)
41. T. Li, Global solutions and zero relaxation limit for a traffic flow model. SIAM J. Appl. Math. **61**, 1042–1061 (2000)
42. T. Li, H. Liu, Stability of a traffic flow model with nonconvex relaxation. Comm. Math. Sci. **3**, 101–118 (2005)
43. M.J. Lighthill, G.B. Whitham, On kinematic waves. II. A theory of traffic flow on long crowded roads. Proc. Roy. Soc. A **229**(1178), 317–345 (1955)
44. T.P. Liu, Hyperbolic conservation laws with relaxation. Comm. Math. Phys. **108**, 153–175 (1987)
45. G. Maruyama, Continuous Markov processes and stochastic equations. Rendiconti del Circolo Matematico di Palermo **4**(1), 48 (1955)
46. K. Nagel, M. Schreckenberg, A cellular automaton model for freeway traffic. J. Phys. I France **2**, 2221–2229 (1992)
47. G.F. Newell, Nonlinear effects in the dynamics of car following. Operations Research **9**, 209–229 (1961)
48. G.F. Newell, A simplified theory of kinematic waves in highway traffic II: Queueing at freeway bottlenecks. Transp. Res. B **27**, 289–303 (1993)
49. P. Noble, Roll-waves in general hyperbolic systems with source terms. SIAM J. Appl. Math. **67**, 1202–1212 (2007)
50. M. Papageorgiou, Some remarks on macroscopic traffic flow modelling. Transp. Res. A **32**, 323–329 (1998)
51. H.J. Payne, Models of freeway traffic and control. Proc. Simul. Counc. **1**, 51–61 (1971)
52. H.J. Payne, FREEFLO: A macroscopic simulation model of freeway traffic. Transp. Res. Rec. **722**, 68–77 (1979)
53. W.F. Phillips, A kinetic model for traffic flow with continuum implications. Transp. Plan. Technol. **5**, 131–138 (1979)
54. L.A. Pipes, An operational analysis of traffic dynamics. J. Appl. Phys. **24**, 274–281 (1953)
55. P.I. Richards, Shock waves on the highway. Operations Research **4**, 42–51 (1956)
56. B. Seibold, M.R. Flynn, A.R. Kasimov, R.R. Rosales, Constructing set-valued fundamental diagrams from jamiton solutions in second order traffic models. Netw. Heterog. Media **8**(3), 745–772 (2013)
57. R.E. Stern, S. Cui, M.L. Delle Monache, R. Bhadani, M. Bunting, M. Churchill, N. Hamilton, R. Haulcy, H. Pohlmann, F. Wu, B. Piccoli, B. Seibold, J. Sprinkle, D.B. Work, Dissipation of stop-and-go waves via control of autonomous vehicles: Field experiments. Transp. Res. C **89**, 205–221 (2018)
58. D.S. Stewart, A.R. Kasimov, State of detonation stability theory and its application to propulsion. J. Propuls. Power **22**(6), 1230–1244 (2006)
59. Y. Sugiyama, M. Fukui, M. Kikuchi, K. Hasebe, A. Nakayama, K. Nishinari, S. Tadaki, S. Yukawa, Traffic jams without bottlenecks – Experimental evidence for the physical mechanism of the formation of a jam. New J. Phys. **10**, 033001 (2008)

60. L.N. Trefethen, Pseudospectra of linear operators. SIAM Rev. **39**(3), 383–406 (1997)
61. L.N. Trefethen, Computation of pseudospectra. Acta Numerica **8**, 247–295 (1999)
62. R. Underwood, Speed, volume, and density relationships: Quality and theory of traffic flow. Technical report, Yale Bureau of Highway Traffic, 1961
63. Y. Wang, M. Papageorgiou, Real-time freeway traffic state estimation based on extended Kalman filter: A general approach. Transp. Res. B **39**, 141–167 (2005)
64. G.B. Whitham, Some comments on wave propagation and shock wave structure with application to magnetohydrodynamics. Comm. Pure Appl. Math. **12**, 113–158 (1959)
65. G.B. Whitham, *Linear and Nonlinear Waves* (Wiley, New York, 1974)
66. D. Work, O.-P. Tossavainen, S. Blandin, A. Bayen, T. Iwuchukwu, K. Tracton, An ensemble Kalman filtering approach to highway traffic estimation using GPS enabled mobile devices, in *47th IEEE Conference on Decision and Control*, pp. 5062–5068, Cancun, Mexico, 2008
67. H.M. Zhang, A non-equilibrium traffic model devoid of gas-like behavior. Transp. Res. B **36**, 275–290 (2002)

Stop-and-Go Waves: A Microscopic and a Macroscopic Description

Caterina Balzotti and Elisa Iacomini

Abstract In this paper we investigate a typical phenomenon of congested traffic: the stop-and-go waves. Since modelling properly this phenomenon is crucial for developing techniques aimed at reducing it, we present two different models: a microscopic and a macroscopic one, both of them able to reproduce stop-and-go waves. In the former, vehicles' dynamics are described by a second-order microscopic Follow-the-Leader model, which is calibrated and validated by real measurements. Data are analysed and compared with the numerical solutions computed by the microscopic model. The latter provides a description of traffic dynamic via the macroscopic second-order CGARZ model. With the numerical implementation, by means of the 2CTM scheme, we test the ability of the model of capturing stop-and-go waves.

1 Introduction

Stop-and-go waves are a typical feature of congested traffic and represent a real danger for drivers. They lead not only to safety hazard, but they also have a negative impact on fuel consumption and pollution.

A stop-and-go wave is detected when vehicles stop and restart without any apparent reason, generating a wave that travels backward with respect to the cars' trajectories [9, 23]. Since modelling properly this phenomenon is crucial for developing techniques aimed at reducing it, a considerable literature is growing up on this topic. This means that a lot of models able to recover stop-and-go waves have been developed in the last years resorting to different approaches.

Starting from the natural idea of tracking every single vehicle, several microscopic models, based on the concept of Follow-the-Leader, grew up for computing positions, velocities and accelerations of each car by means of systems of ordinary

C. Balzotti (✉) · E. Iacomini
SBAI Department, Sapienza Università di Roma, Rome, Italy
e-mail: caterina.balzotti@sbai.uniroma1.it; elisa.iacomini@sbai.uniroma1.it

G. Puppo, A. Tosin (eds.), *Mathematical Descriptions of Traffic Flow: Micro, Macro and Kinetic Models*, SEMA SIMAI Springer Series 12,
https://doi.org/10.1007/978-3-030-66560-9_4

differential equations (ODEs) [9, 23]. Other ways go from kinetic [11, 12] to macroscopic fluid-dynamic [3, 27] approach, focusing on averaged quantities, such as the traffic density and the speed of the traffic flow, by means of systems of hyperbolic partial differential equations (PDEs), in particular conservation laws. In this way we lose the detailed level of vehicles' description, indeed they become indistinguishable from each other. The choice of the scale of observation mainly depends on the number of the involved vehicles, the size of the network and so on.

Moreover, several real experiments took place, just see [32], where the dynamics of pedestrians, cyclists and drivers are described in a unique framework, or [25], where an autonomous vehicle is used to reduce the stop-and-go waves. Furthermore autonomous vehicles could be trained to avoid triggering stop-and-go waves by reducing the acceleration rate or influencing the reaction time.

Real traffic data play a fundamental role in this framework; indeed, detecting and observing real stop-and-go waves points out the main features of this phenomenon. Starting from the huge literature we mentioned above, we develop a new second-order microscopic model specially conceived to study and reproduce stop-and-go waves. Due to its simplicity, we are able to calibrate and validate this model from a set of real data coming from Autovie Venete S.p.A. (AV), an Italian company operating on the highways of the North-East of Italy. We estimate the parameters with a trial-and-error method, fitting the right velocity of the backward propagation of the wave.

On the other hand microscopic models have well-known drawbacks. Even if they lead to more precise and accurate description due to the natural way of tracking each single vehicle and following its trajectory, they become quickly unfeasible as the number of vehicles grows enough or if the considered road network is large. Indeed managing all the positions and the velocities at each time is expensive in terms of memory and time. Such computational limitations are naturally overcome with the use of a macroscopic description.

The investigation on stop-and-go waves from a macroscopic point of view is a topic of growing interest. As explained in [17], the main components of traffic oscillations are their formation and propagation. First-order models for traffic flow, such as the well-known Lighthill–Whitham–Richards (LWR) model [19, 22], are able to simulate the formation of traffic waves, but they are not sufficient to study their propagation and they cannot reproduce traffic instabilities due to initial perturbations, as observed in [24]. Another relevant drawback of first-order models is that they produce infeasible solutions with unbounded accelerations.

Second-order models have been introduced to overcome the limits of first-order models. We cite, in particular, the pioneer papers by Aw, Rascle and Zhang [1, 31], and subsequent generalizations such as [4, 7, 18]. Traffic models of second order turned out to be more suitable to reproduce complex traffic phenomena, in particular the formation and propagation of stop-and-go waves (see for instance [10, 15, 20]). Moreover, many studies pointed out that traffic instabilities can be captured by taking into account different driver behaviours, e.g., timid and aggressive drivers. In [30] a multi-class LWR model is proposed, and it is shown that such a model is able to reproduce many traffic phenomena that cannot be caught by a simple LWR model. The importance of different driver behaviours has been further investigated in [16, 21].

The macroscopic model employed in this work is specifically designed to be calibrated with real data in order to get realistic simulations. The validation of this model with the dataset mentioned above is still under investigation, but, as a first attempt, we show that it is able to reproduce stop-and-go waves and therefore it deserves further studies.

1.1 Paper Organization

In Sect. 2 we propose a new microscopic model derived from the model recently introduced by Zhao and Zhang [32] to reproduce stop-and-go waves. We calibrate and validate the model using real traffic data coming from Autovie Venete S.p.A. (AV). Section 3 is devoted to the study of stop-and-go waves with the macroscopic Collapsed Generalized Aw–Rascle–Zhang model [8], showing some numerical tests able to capture stop-and-go waves. We conclude the paper with some comments and an overview on future directions.

1.2 Goal

The aim of this work is developing models able to reproduce stop-and-go waves, from both microscopic and macroscopic points of view. Since the choice of the scale of observation mainly depends on the number of the involved vehicles, the size of the network and so on, we provide suitable tools in both scenarios.

Moreover, real data play a crucial role in this framework. Validation and calibration of the microscopic model is one of the main novelties of this work.

2 Microscopic Model

In this section we describe a new microscopic second-order model specifically conceived to reproduce stop-and-go waves. This model is nothing but a minimal version of the model recently introduced by Zhao and Zhang [32] to describe the dynamics of vehicles, bicycles and pedestrians in a unified framework. Our model is 'minimal' in the sense that it is obtained from the Zhao and Zhang's model dropping all the terms, one by one, which are not strictly necessary to reproduce realistic stop-and-go waves.

Assume that N vehicles (of length $\ell_N > 0$) are moving along a single-lane infinite road where overtaking is not possible. This means that cars are ordered and have to follow the first one that is called the *leader*. Cars that are not the leader are

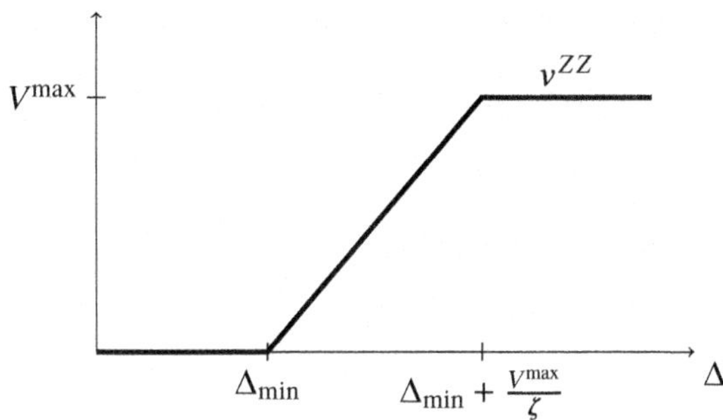

Fig. 1 Profile of the $v^{ZZ}(\Delta)$ function (3)

termed *followers*. Let $X_k(t)$ be the position of the k-th car at time $t > 0$ and $V_k(t)$ its instantaneous velocity, for $k = 0, \ldots, N$.

The dynamics is governed by the distance between adjacent vehicles and described by a system of ordinary differential equations (ODEs), as a typical *Follow-the-Leader* second-order microscopic model [2, 5, 28]:

$$\begin{cases} \dot{X}_k(t) = V_k(t), & k \leq N \\ \dot{V}_k(t) = A(X_k(t), X_{k+1}(t), V_k(t), V_{k+1}(t)), & k < N \\ \dot{V}_N(t) = 0. \end{cases} \tag{1}$$

Our choice for the acceleration term is the following:

$$A\big(X, X', V, V'; (\tau, \zeta, \Delta_{\min}, V_{\max})\big) = \frac{1}{\tau}\Big(v^{ZZ}\big(X' - X\big) - V\Big), \tag{2}$$

where

$$v^{ZZ}(\Delta) := \begin{cases} 0, & \Delta \leq \Delta_{\min}, \\ \zeta(\Delta - \Delta_{\min}), & \Delta_{\min} \leq \Delta \leq \Delta_{\min} + V_{\max}/\zeta, \\ V_{\max}, & \Delta \geq \Delta_{\min} + V_{\max}/\zeta. \end{cases} \tag{3}$$

We can consider v^{ZZ} as the desired velocity a vehicle would achieve, while τ models the time reaction of drivers to align their velocity to the desired one, following [2] where only the relaxation term is different from 0 (Fig. 1).

Moreover $\zeta > 0$ is a parameter and $\Delta_{\min} > \ell_N$ is the minimum critical spacing distance between the centres of mass of a vehicle and the preceding one. Note that this minimal model, i.e. this reduced version, unlike the original one [32], is deterministic. Moreover, one should note that the condition $X_{k+1}(0) - X_k(0) \geq \ell_N \Rightarrow X_{k+1}(t) - X_k(t) \geq \ell_N \; \forall t$ is not a priori guaranteed.[1]

[1]The question arises why this condition should hold true in the context of traffic modelling, considering the fact that rear-end collisions are actually possible in real life.

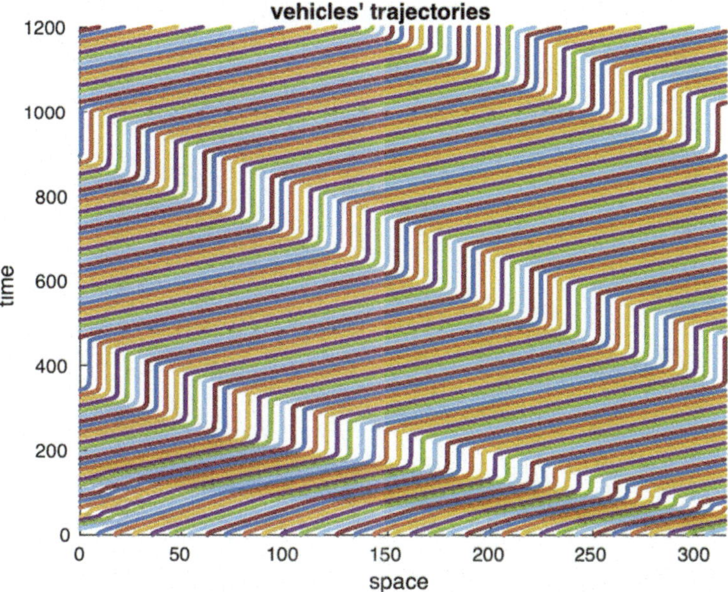

Fig. 2 Space-time trajectories of vehicles obeying to the system (1)–(2)–(3) with $N = 34$, $\alpha = 0.6$, $\Delta_{\min} = 7.89$, $V_{\max} = 1$, $\tau = 4.86$ and $L = 314$

Remark 1 One could also consider different values for ℓ_k in order to model, for example, different types of vehicles. For simplicity, in the following, ℓ_k is assumed to be the same for each k, i.e. all the vehicles are equal.

In Figs. 2 and 3 we show a typical solution to the system (1)–(2)–(3) in the case of a *circular road* of length L. Initial conditions are $X_k(0) = \frac{kL}{N+1}$ and $V_k(0) = 0$, for $k = 1, \ldots, N$. Numerical integration is obtained by the explicit Euler scheme on a road segment $[0, L]$ with periodic boundary conditions.

It can be seen that backward stop-and-go waves are immediately generated by the small perturbation in the initial positions of the vehicles. Indeed looking from the bottom of the right side in Fig. 2, vehicles start to stop after a few time steps, i.e. vertical lines mean that vehicles are in the same positions while time is changing, and these perturbations are increasing and travelling backwards. Note that small perturbations in the initial velocity lead to similar effects as well. Vehicles are initially equi-spaced

$$X_k(0) - X_{k-1}(0) = \frac{L}{N+1}, \quad k > 1$$

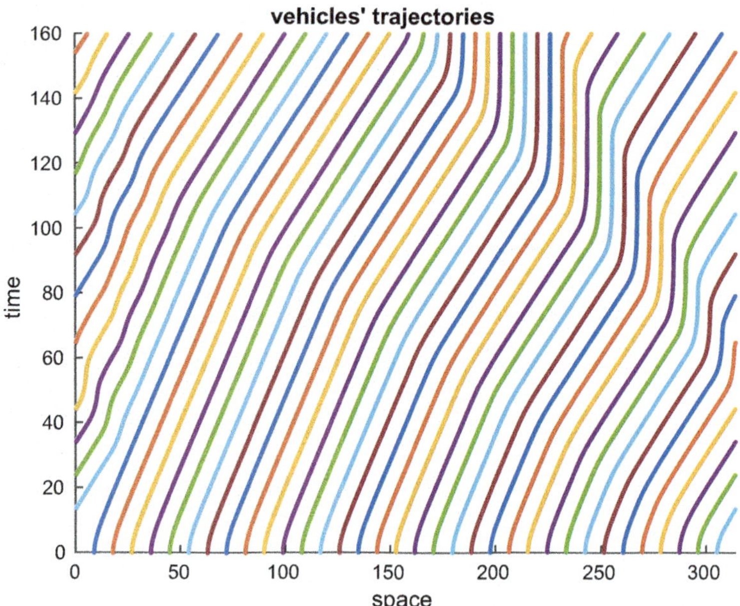

Fig. 3 Zoom of the trajectories shown in Fig. 2 around initial time. It is well-visible the emergence of the stop-and-go wave from the interaction between the first and the last vehicle

with the exception of the couple $(N,1)$ (first vehicle in X_1 is just in front of the N-th vehicle in X_N because of the periodic boundary conditions), for which we have

$$X_1(0) + L - X_N(0) = \frac{L}{N+1} + L - \frac{NL}{N+1} = \frac{2L}{N+1}.$$

2.1 *Real Data Validation*

Let us focus on real data coming from Autovie Venete S.p.A. (AV), an Italian company operating on the highways of the North-East of Italy, which has several sensors and radars to supervise and control traffic evolution all over its road network.

For example, let us analyse data collected on 21.09.2018 along the east direction of the highway called A4, from Venice to Trieste. Starting from a small slowdown located at 489 km at time 12:10, a little queue is formed 10 min later between 485 and 486 km. From AV reports, we know that the queue grows and moves backwards: at time 12:30, the queue is between 482 and 483 km and 1 h later is between 463 and 468 km, as shown in Fig. 4.

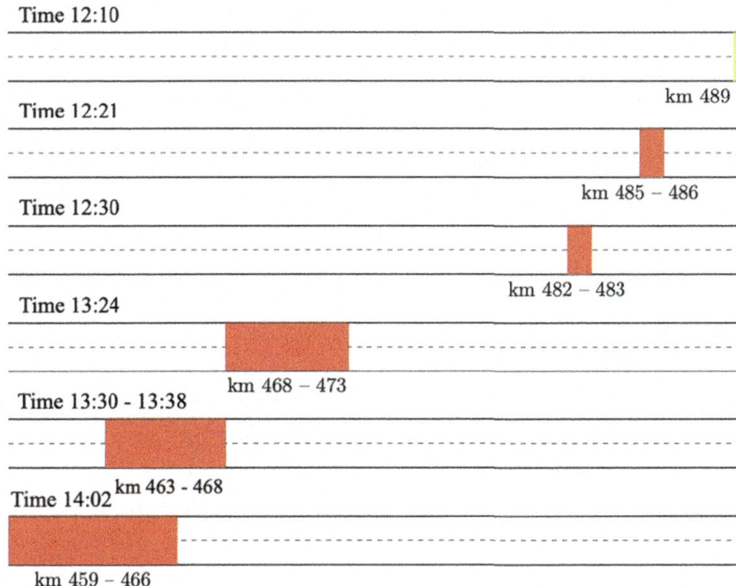

Fig. 4 Real data on the queue on the A4 highway on 21.09.2018

This is the typical stop-and-go wave behaviour: the initial perturbation increases and propagates backwards, so vehicles begin to stop and restart even far from the origin of perturbation itself.

Looking at the data from fixed sensors, let us focus on the flux of vehicles. We can easily observe that flux data are oscillating a lot, as the red line in Fig. 5 suggests. Moreover, there are minutes in which only a few vehicles are detected, and this means either that a few vehicles travel along the highway or that vehicles are slow (or stopped), i.e. the road is congested. In order to distinguish the two cases, we consider the Gaussian mean, the black line in Fig. 5. Considering the sensors on the East direction of A4 at 466.7 km from 12:00 to 16:00 on 21.09.2018 as before, we note that the flux detected, in particular the Gaussian mean, is almost constant until 13:30. From 13:30 to 14:00, the flux is almost 0, as we expected seeing Fig. 4. Moreover from 15:00 to 15:30, another queue is detected.

Now let us consider a straight road in order to simulate the trajectories reported in AV real data. Starting with an empty road, we use the flux data coming from sensors in order to recover left boundary condition. These measurements are referring to the stop-and-go wave detected on 21.09.2018 on the East direction of A4, as above, see Figs. 4 and 5.

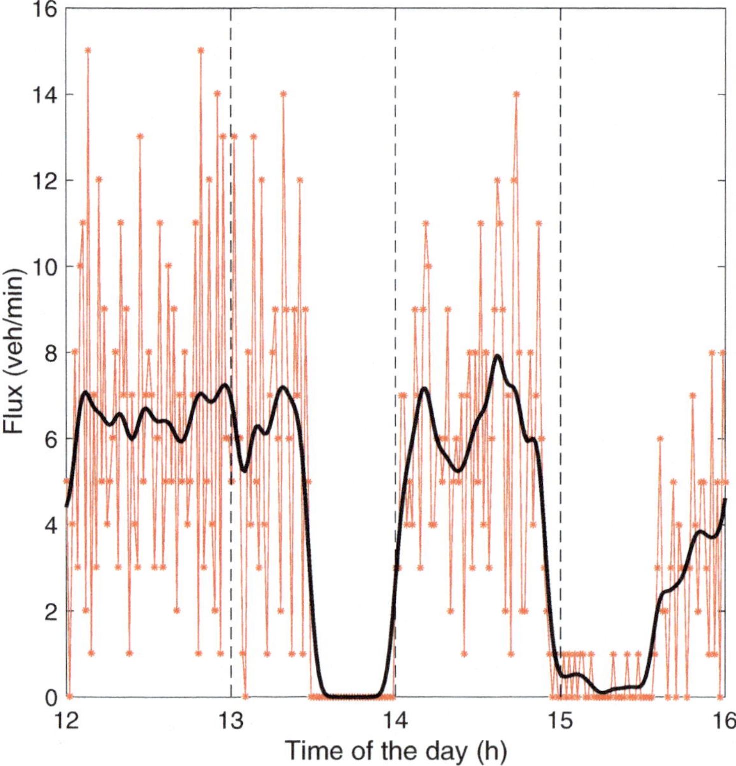

Fig. 5 Flux of heavy vehicles on A4 highway on 21.09.2018 (red line) and its Gaussian mean (black line)

Let us modify the model (2)–(3) considering two different reaction times: one for accelerations and the other one for decelerations. The system becomes as follows:

$$\begin{cases} \dot{X}_k(t) = V_k(t), \\ \begin{cases} \dot{V}_k(t) = \frac{1}{\tau_A}\left(v^{ZZ}\left(X_{k+1}(t) - X_k(t)\right) - V_k(t)\right) & \text{if } v^{ZZ} - V_k > 0, \\ \dot{V}_k(t) = \frac{1}{\tau_F}\left(v^{ZZ}\left(X_{k+1}(t) - X_k(t)\right) - V_k(t)\right) & \text{if } v^{ZZ} - V_k < 0, \end{cases} \end{cases} \tag{4}$$

where τ_A and τ_F are the acceleration/deceleration reaction times, respectively. This choice is motivated by the fact that heavy vehicles' reaction to accelerations and decelerations is different in time, i.e. braking must be more efficient (also for safety) than acceleration, see [13, 14, 26].

We modify also the function v^{ZZ}, introducing a discontinuity in its profile as in Fig. 6. After duly calibration, the solution obtained with this microscopic model is

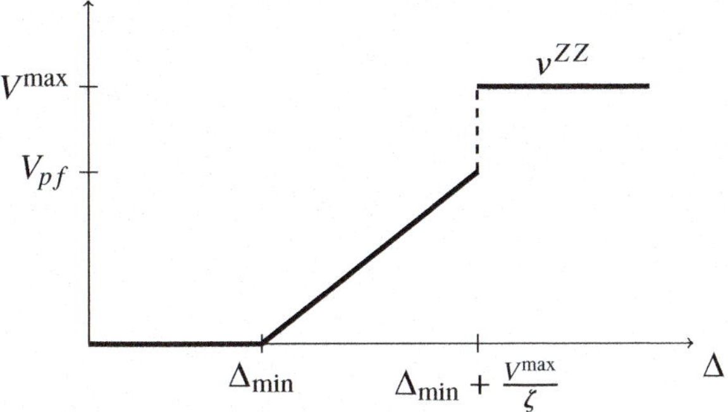

Fig. 6 Profile of the calibrated v^{ZZ} function

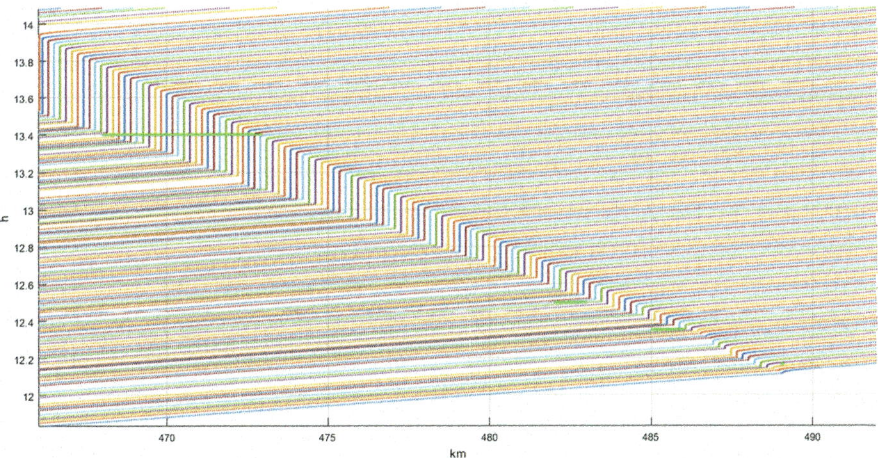

Fig. 7 Heavy vehicles' trajectories simulated with AV report measurements overlapping (green lines)

really close to real data as shown in Fig. 7, where the light green lines stand for the length of the queue recorded in AV reports.

3 Macroscopic Model

In this section we investigate the stop-and-go waves from a macroscopic point of view. We propose a second-order traffic flow model, which takes into account different driver behaviours by means of a family of fundamental diagrams.

3.1 CGARZ Model

The macroscopic second-order traffic model used to study the stop-and-go waves is the Collapsed Generalized Aw–Rascle–Zhang (CGARZ) model [8], which belongs to the family of Generic Second-Order Models (GSOM) [6, 18]. The models of this family are described by

$$\begin{cases} \rho_t + (\rho v)_x = 0, \\ w_t + v w_x = 0, \end{cases} \qquad (5)$$
$$\text{with } v = V(\rho, w),$$

for a proper velocity function V. Here, $\rho = \rho(x,t)$ is the density of vehicles, $v = v(x,t)$ is the velocity and $w = w(x,t)$ is a property of drivers, which is advected with density. By introducing the total property $y = \rho w$, we rewrite the problem (5) in conservative form as

$$\begin{cases} \rho_t + (\rho v)_x = 0 \\ y_t + (y v)_x = 0 \end{cases} \qquad (6)$$
$$\text{with } v = V\Big(\rho, \frac{y}{\rho}\Big).$$

System (6) is more suitable for the numerical implementation in Sect. 3.3. The GSOM are characterized by a family of flux functions $Q(\rho, w) = \rho V(\rho, w)$, and the role of w is to distinguish different driver behaviours by choosing the fundamental diagram from the family $Q(\rho, w)$.

The need to use a family of fundamental diagrams instead of a single flux function has been suggested by real data. Indeed, real data show that at fixed density value does not correspond a unique flux value but multiple ones, as we can see in Fig. 8. Such data are obtained from the NSGIM dataset [29].

The idea of the CGARZ model is that the property w does not affect the flux during the free-flow phase, but only during the congested phase. Thus,

$$Q(\rho, w) = \begin{cases} Q_f(\rho) & \text{if } 0 \le \rho \le \rho_f \\ Q_c(\rho, w) & \text{if } \rho_f \le \rho \le \rho^{\max}, \end{cases} \qquad (7)$$

for suitable functions Q_f and Q_c. Here ρ_f is the density threshold between the two traffic flow phases and $\rho^{\max}$ the maximum density. Each curve of the family of fundamental diagrams must be a strictly concave function in C^1 that is equal to 0 both when $\rho = 0$ and $\rho = \rho^{\max}$, for each w, and that is increasing in the second argument, i.e. $\frac{\partial Q(\rho, w)}{\partial w} > 0$ for $\rho_f < \rho < \rho^{\max}$. In Fig. 9 we show an example of a family of flux and speed functions fitting in the framework (7).

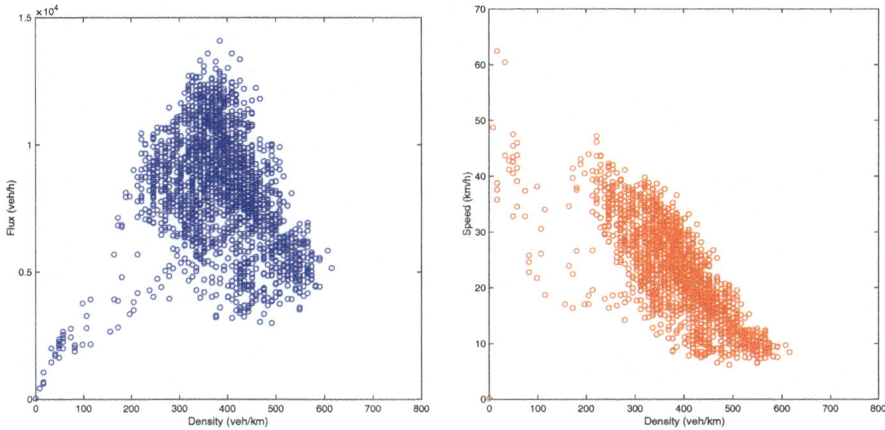

Fig. 8 Flow-density data (left) and speed-density data (right) derived from the NGSIM dataset

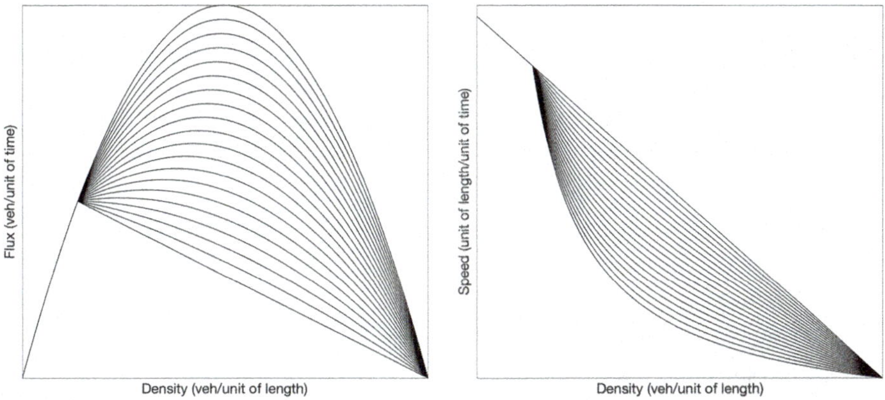

Fig. 9 Family of flux functions $Q(\rho, w)$ (left) and family of speed functions $V(\rho, w)$ (right)

Remark 2 The domain of w is a proper interval $[w_L, w_R]$, which depends on the choice of the family of flux functions $Q(\rho, w)$. Note that, since Q is increasing with respect to w, small values of w are associated to timid drivers who send a lower flux of vehicles, while high values of w are associated to aggressive drivers with higher fluxes.

3.2 Generating Stop-and-go Waves

In order to generate stop-and-go waves we modify problem (5) by adding a source term to the equation for w, i.e.

$$\begin{cases} \rho_t + (\rho v)_x = 0, \\ w_t + v w_x = \dfrac{Q_{eq} - Q}{\tau_1} + \dfrac{\delta}{\tau_2} w_x, \end{cases} \tag{8}$$

where $Q_{eq} = Q(\rho, w_{eq})$ for a certain w_{eq}, $Q = Q(\rho, w)$, τ_1 is a relaxation time and τ_2 is the time needed to backward propagate information to a distance δ. The parameter w_{eq} is an equilibrium for the variable w, and it represents the "average" behaviour of drivers. This means that, if w_{eq} is close to w_L, then we are looking for perturbations with respect to the case of quiet drivers, if w_{eq} is close to w_R, then we are looking for perturbations with respect to the case of aggressive drivers. Problem (8) can be written in conservative form as

$$\begin{cases} \rho_t + (\rho v)_x = 0, \\ y_t + (y v)_x = \dfrac{Q_{eq} - Q}{\tau_1} \rho + c_0 y_x - c_0 \dfrac{y}{\rho} \rho_x, \end{cases} \tag{9}$$

where $c_0 = \dfrac{\delta}{\tau_2}$.

The idea is that traffic delays are associated to lower values of w, and thus to a reduction of the flux of vehicles. If we perturb w, we generate a queue and the source term allows to take into account what happens forward on w and, as a consequence, the queue travels backwards.

3.3 Numerical Implementation

Let us consider the traffic model (9) on a single-lane road $[0, L]$ during a time interval $[0, T]$. We discretize space and time via a grid of $N_x \times N_t$ cells of length $\Delta x \times \Delta t$. The numerical scheme used in this work is the Second-Order Cell Transmission Model (2CTM) [8] with which we estimate $\rho_i^{n+1} = \rho(x_i, t^n)$ and $y_i^{n+1} = y(x_i, t^{n+1})$ as

$$\begin{aligned} \rho_i^{n+1} &= \rho_i^n - \frac{\Delta t}{\Delta x}\left(F^{\rho,n}_{i+\frac{1}{2}} - F^{\rho,n}_{i-\frac{1}{2}}\right), \\ y_i^{n+1} &= y_i^n - \frac{\Delta t}{\Delta x}\left(w_{i+1}^n F^{\rho,n}_{i+\frac{1}{2}} - w_i^n F^{\rho,n}_{i-\frac{1}{2}}\right) + \frac{\Delta t}{\tau_1} \rho_i^n \left(Q_{eq}(\rho_i^n) - Q(\rho_i^n, w_i^n)\right) \\ &\quad + \frac{c_0 \Delta t}{\Delta x}(y_{i+1}^n - y_i^n) - \frac{c_0 \Delta t}{\Delta x} w_i^n (\rho_{i+1}^n - \rho_i^n), \end{aligned}$$

where $F^{\rho,n}_{i\pm\frac{1}{2}}$ are the numerical fluxes. We refer to [8] for further details on the numerical scheme.

Starting from an initial state $(\rho_0(x), w_0(x))$, we perturb the property w for a certain time t on the node x_κ, by reducing it to its minimum value w_L. The parameters that describe the test are $L = 30$, $T = 0.4$, $\Delta x = 0.1$, $\Delta t = 4 \times 10^{-4}$, $t = 3 \times 10^{-3}$, $x_\kappa = 25$ and $V_{\max} = 130$. The maximum density $\rho^{\max}$ is a property of the road, defined as $\rho^{\max} =$ number of lanes/length of vehicles. Assuming that each vehicle has a length of 7.5 m, we have $\rho^{\max} = 133$ veh/km. The threshold parameter ρ_f is chosen as $\rho^{\max}/7$. The interval for the property w is $[2.1 \times 10^3, 4.3 \times 10^3]$, while the parameters for the source term are $\tau_1 = 2.6$, $\tau_2 = 2.6 \times 10^{-3}$, $\delta = 0.2$ and $w_{eq} = w_R$. Finally, the initial conditions are $\rho_0(x) = 70$ and $w_0(x) = w_R$.

In Fig. 10 we show the formation of the wave and its backward propagation. The perturbation on the property w causes the formation of the wave, and it travels backward, thanks to the source term that allows w to "look forward" along the road. Meanwhile the relaxation term τ_1 allows the property w to restore its equilibrium

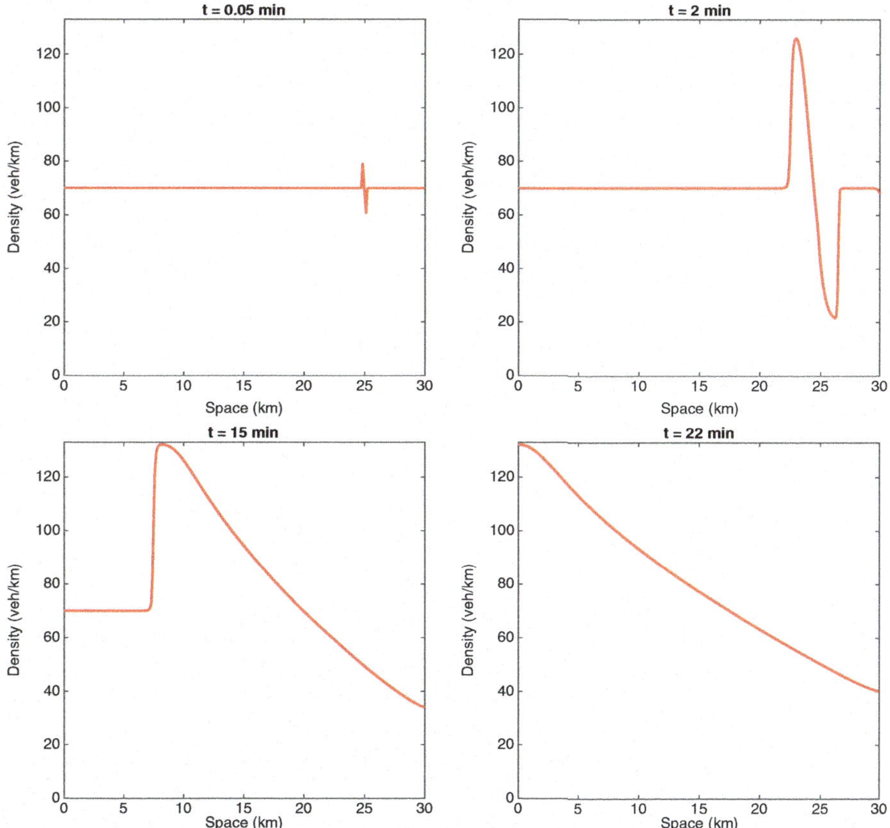

Fig. 10 Backward propagation of the stop-and-go wave at different times

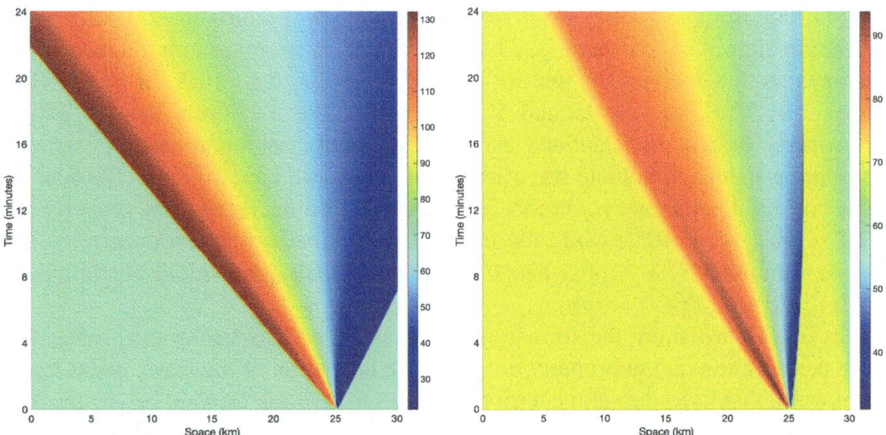

Fig. 11 Density of vehicles in space and time when $w_0 = w_R$ (left) and $w_0 = w_M$ (right)

value w_{eq}. We observe that the width of the queue increases going backward in space.

In Fig. 11 we show the 3D plots of the density of vehicles in the whole road at each time of the simulation when $w_0 = w_R$ and $w_0 = \frac{w_L + w_R}{2} =: w_M$. From the plot on the left, it is clear that the width of the queue increases going backward in space, thus it involves more vehicles over time. The plot on the right, instead, shows a smaller queue than the one on the left, since the maximum density reached is 90 veh/km, which is lower than $\rho^{\max}$. Moreover, the height of the queue decreases in time. The difference between these two examples is due to the initial behaviour of drivers: when $w_0 = w_R$, all drivers are aggressive and they move with maximum flux so that they supply the queue; when $w_0 = w_M$, instead, the drivers have a quieter behaviour, which prevents the amplification of the queue.

4 Conclusions

In this work we proposed a microscopic and a macroscopic second-order traffic flow model able to reproduce stop-and-go waves, showing simulations where stop-and-go waves are well-recognizable. After detecting stop-and-go waves from real data, we investigated this phenomenon from both points of view, since an accurate description requires to overcome the issues coming from single scale models.

We developed an easy-to-use second-order microscopic Follow-the-Leader model, calibrated and validated by real measurements coming from the collaboration with Autovie Venete. We found out that real data well-fitted with the simulations obtained with the microscopic model. On the other hand, we tested the second-order CGARZ model, combined with a proper source term, showing that it is able to simulate the formation and propagation of stop-and-go waves.

There are several open problems related to the work presented here. First of all, analogously to what has been already done with the microscopic model, future work will aim at calibrating the macroscopic model with real data, in order to compare the simulated results with the observed phenomena.

Starting from models able to reproduce stop-and-go waves, another research direction could be to control the vehicles' dynamics in order to avoid the triggering of this phenomenon, for example with the help of trained autonomous vehicles as in [25]. Just to begin, we can associate a control problem to the microscopic model that is easier to deal with. The control can be done on the acceleration term by means of the desired velocity or by means of the reaction time.

Finally, in order to compare the results obtained with the two different observation scales, one could recover the many particles' limit of the proposed microscopic model and make the comparison with the resulting macroscopic model. Moreover, the microscopic counterpart of the macroscopic model presented in this work could also be investigated.

Acknowledgments Authors want to thank Maya Briani and Emiliano Cristiani for the useful suggestions and comments on this work.

References

1. A. Aw, M. Rascle, Resurrection of "second order" models of traffic flow. SIAM J. Appl. Math. **60**(3), 916–938 (2000)
2. A. Aw, A. Klar, M. Rascle, T. Materne, Derivation of continuum flow traffic models from microscopic follow-the-leader models. SIAM J. Appl. Math. **63**, 259–278 (2002)
3. R. Borsche, A. Klar, A nonlinear discrete velocity relaxation model for traffic flow. SIAM J. Appl. Math. **78**(5), 2891–2917 (2018)
4. R.M. Colombo, Hyperbolic phase transitions in traffic flow. SIAM J. Appl. Math. **63**(2), 708–721 (2003)
5. M. Di Francesco, S. Fagioli, M.D. Rosini, Many particle approximation of the Aw-Rascle-Zhang second order model for vehicular traffic. Math. Biosci. Eng. **14**(1), 127–141 (2017). DOI 10.3934/mbe.2017009
6. S. Fan, *Data-Fitted Generic Second Order Macroscopic Traffic Flow Models*. Ph.D. thesis, (Temple University, Philadelphia, 2013)
7. S. Fan, M. Herty, B. Seibold, Comparative model accuracy of a data-fitted generalized Aw-Rascle-Zhang model. Netw. Heterog. Media **9**(2), 239–268 (2014)
8. S. Fan, Y. Sun, B. Piccoli, B. Seibold, D.B. Work, *A Collapsed Generalized Aw-Rascle-Zhang Model and Its Model Accuracy* (2017). arXiv preprint arXiv:1702.03624
9. M.R. Flynn, A.R. Kasimov, J.C. Nave, R.R. Rosales, B. Seibold, Self-sustained nonlinear waves in traffic flow. Phys. Rev. E **79**, 56–113 (2009)
10. D. Helbing, A. Hennecke, V. Shvetsov, M. Treiber, MASTER: macroscopic traffic simulation based on a gas-kinetic, non-local traffic model. Transport. Res. B-Meth. **35**(2), 183–211 (2001)
11. M. Herty, R. Illner, On stop-and-go waves in dense traffic. Kinet. Relat. Mod. **1**(3), 437–452 (2008)
12. M. Herty, G. Puppo, S. Roncoroni, G. Visconti, The BGK approximation of kinetic models for traffic. Kinet. Relat. Mod. **13**(2), 279–307 (2020)

13. B.S. Kerner, *The Physics of Traffic: Empirical Freeway Pattern Features, Engineering Applications, and Theory* (Springer, Berlin, 2012)
14. B.S. Kerner, S.L. Klenov, Deterministic microscopic three-phase traffic flow models. J. Phys. A-Math. Gen. **39**(8), 1775 (2006)
15. B. Kerner, P. Konhäuser, M. Schilke, Deterministic spontaneous appearance of traffic jams in slightly inhomogeneous traffic flow. Phys. Rev. E **51**(6), 6243 (1995)
16. J.A. Laval, Hysteresis in traffic flow revisited: an improved measurement method. Transport. Res. B-Meth. **45**(2), 385–391 (2011)
17. J.A. Laval, L. Leclercq, A mechanism to describe the formation and propagation of stop-and-go waves in congested freeway traffic. Philos. T. R. Soc. A **368**(1928), 4519–4541 (2010)
18. J.P. Lebacque, S. Mammar, H.H. Salem, Generic second order traffic flow modelling, in *Transportation and Traffic Theory 2007*. Papers Selected for Presentation at ISTTT17 (2007), pp. 755–776
19. M.J. Lighthill, G.B. Whitham, On kinematic waves II. A theory of traffic flow on long crowded roads. P. R. Soc. Lond. A Mat. **229**(1178), 317–345 (1955)
20. D. Ngoduy, Application of gas-kinetic theory to modelling mixed traffic of manual and ACC vehicles. Transportmetrica **8**(1), 43–60 (2012)
21. D. Ngoduy, Effect of driver behaviours on the formation and dissipation of traffic flow instabilities. Nonlinear Dynam. **69**(3), 969–975 (2012)
22. P.I. Richards, Shock waves on the highway. Oper. Res. **4**, 42–51 (1956)
23. B.G. Ros, V.L. Knoop, B. van Arem, S.P. Hoogendoorn, Empirical analysis of the causes of stop-and-go waves at sags. IET Intell. Transp. Syst. **8**(5), 499–506 (2014)
24. M. Schönhof, D. Helbing, Empirical features of congested traffic states and their implications for traffic modeling. Transport. Sci. **41**(2), 135–166 (2007)
25. R.E. Stern, S. Cui, M.L. Delle Monache, R. Bhadani, M. Bunting, M. Churchill, N. Hamilton, R. Haulcy, H. Pohlmann, F. Wu, B. Piccoli, B. Seibold, J. Sprinkle, D.B. Work, Dissipation of stop-and-go waves via control of autonomous vehicles: field experiments. Transport. Res. Part C-Emer. **89**, 205–221 (2018). DOI 10.1016/j.trc.2018.02.005
26. T. Toledo, Driving behaviour: models and challenges. Transport Rev. **27**(1), 65–84 (2007)
27. A. Tordeux, G. Costeseque, M. Herty, A. Seyfried, From traffic and pedestrian follow-the-leader models with reaction time to first order convection-diffusion flow models. SIAM J. Appl. Math. **78**(1), 63–79 (2018)
28. M. Treiber, A. Kesting, Traffic flow dynamics, in *Traffic Flow Dynamics: Data, Models and Simulation* (Springer, Berlin, 2013)
29. US Department of Transportation, Federal Highway Administration: Next Generation Simulation (NGSIM). http://ops.fhwa.dot.gov/trafficanalysistools/ngsim.htm
30. G. Wong, S. Wong, A multi-class traffic flow model–an extension of LWR model with heterogeneous drivers. Transport. Res. A-Pol. **36**(9), 827–841 (2002)
31. H.M. Zhang, A non-equilibrium traffic model devoid of gas-like behavior. Transport. Res. B-Meth. **36**(3), 275–290 (2002)
32. Y. Zhao, H.M. Zhang, A unified follow-the-leader model for vehicle, bicycle and pedestrian traffic. Transport. Res. B-Meth. **105**, 315–327 (2017)

An Overview of Non-local Traffic Flow Models

Felisia Angela Chiarello

Abstract We give an overview of mathematical traffic flow models with non-local velocity. More precisely, we consider conservation laws with flux functions depending on an integral evaluation of the density of vehicles through a convolution product. We summarize the analytical results recently obtained for this kind of models, and we provide some numerical simulations illustrating the behaviour of different groups of drivers or vehicles.

1 Introduction

There are three different important approaches to model traffic flow phenomena: microscopic models, mesoscopic models and macroscopic models. Microscopic models are based on the individual vehicles and their interactions; in particular, micro-models are cellular automata that prescribe stochastic rules how vehicles behave into cells in which the road is divided; mesoscopic models apply gas-kinetic equations to describe the traffic flow; and macroscopic models consider quantities that characterize the collective behaviour of vehicles. In this paper we will consider the framework of macroscopic models. The first macroscopic traffic flow model, based on fluid-dynamics equations, has been introduced in the transportation literature since the mid-fifties of last century. It is the Lighthill, Whitham and Richards (LWR) model [27, 29]. The LWR model consists in one scalar equation that expresses the conservation of the number of cars:

$$\partial_t \rho + \partial_x (\rho\, v(\rho)) = 0,$$

with $\rho = \rho(t, x)$ representing the mean traffic density, i.e. the number of vehicles per unit length and v denoting the mean velocity. Due to the difficulty to simulate

F. A. Chiarello (✉)
Inria Sophia Antipolis Mediterranée, Université Côte d'Azur, Sophia Antipolis, France
e-mail: felisia.chiarello@inria.fr

G. Puppo, A. Tosin (eds.), *Mathematical Descriptions of Traffic Flow: Micro, Macro and Kinetic Models*, SEMA SIMAI Springer Series 12,
https://doi.org/10.1007/978-3-030-66560-9_5

the formation of traffic jams, several approaches have been developed during the years, addressing the need for more sophisticated models. In particular, the classical LWR model does not match the experimental data because it is based on the assumption that the mean traffic velocity is a function of the traffic density, which is not realistic in congested regimes. For these reasons, high-order models were developed. In particular, in [28, 31] and [3, 32], the authors considered second-order models consisting in a mass conservation equation for the density and an acceleration balance law for the speed. Recently, "non-local" versions of the LWR model have been proposed in [7, 30]. In this type of models, the speed depends on a weighted mean of the downstream traffic density. As a consequence, the speed is a Lipschitz function with respect to space and time variables, ensuring bounded acceleration and overcoming the limitation of classical macroscopic models that allows for speed discontinuities. Non-local traffic models are intended to describe the behaviour of drivers that adapt their velocity with respect to what happens in front of them. For this reason, the flux function depends on a "downstream" convolution term between the density of vehicles and a kernel function supported on the negative axis $]-\infty, 0]$. As in classical (local) models, the speed is a monotone non-increasing function, because the higher is the density of cars on a road, the lower is their speed. There are general existence and uniqueness results for non-local equations in [2, 21] for the scalar case in one-space dimension, in [15, 23] for multi-dimensional scalar equations and in [1] for the multi-dimensional system case. To prove the existence of solutions for these non-local models, there are mainly two different approaches. One is providing suitable compactness estimates on a sequence of approximate solutions constructed through finite volume schemes, as in [1, 2]. Another approach relies on characteristics and fixed-point theorems, as proposed in [21, 23]. The first approach [2] requires Kružkov-type entropy conditions to prove the $\mathbf{L^1}$-stability with respect to the initial data through the doubling of variable technique, while in [21, 23] the uniqueness of weak solutions is obtained directly from the fixed point theorem. An interesting point is the numerical approximation of the solutions of non-local models. It results challenging due to the high non-linearity of the system and the dependence of the flux function on integral terms, which highly increase the computational cost. There are different ways to numerically integrate non-local conservation laws. A first-order Lax–Friedrichs-type numerical scheme is used to approximate the problem and to prove the existence of solutions in [1, 2, 6, 7, 20]. Another first-order numerical scheme is proposed in [19], where an upwind numerical scheme for a class of non-local flux problems is presented. Alternative first-order schemes based on the splitting of the non-local equation into two equations, i.e. the Lagrangian and the remap steps are proposed in [14]. In comparison with Lax–Friedrichs scheme and upwind scheme, Lagrangian-Antidiffusive Remap schemes are much less diffusive. Concerning high-order numerical schemes, it is worth citing the papers [8, 13, 18]. In [8], the authors propose discontinuous Galerkin and finite volume WENO schemes to obtain high-order approximations of non-local scalar conservation laws in one space dimension, where the velocity function depends on a weighted mean of the conserved quantity. The discontinuous Galerkin schemes give the best results, but under a very restrictive Courant–Friedrichs–Lewy (CFL) condition. On the contrary, finite volume WENO schemes can be implemented on larger time steps. In [13], the authors extend to 1D systems the finite volume WENO schemes proposed in [8]. In [18], central WENO schemes are proposed, which, in contrast to the other high-order schemes for non-local conservation laws, neither require a restrictive CFL condition nor an additional reconstruction step. The paper is organized as follows: Sects. 1–4 consist in a review from existing papers. In particular, Sect. 2 is devoted to scalar non-local traffic flow models and the corresponding analytical results. A non-local multi-class traffic flow model is recalled in Sect. 3, and Sect. 4 presents an upwind numerical scheme, used in Sect. 5 for numerical tests of the multi-class non-local model with three classes of vehicles. These numerical results are the novelty of this paper.

2 A Class of Scalar Non-local Traffic Flow Models

In this section, we recall the class of scalar conservation laws with non-local flux arising in traffic modelling studied in [9]

$$\partial_t \rho + \partial_x \left(f(\rho) v(\omega_\eta * \rho) \right) = 0, \qquad x \in \mathbb{R},\ t > 0, \tag{1}$$

where

$$\omega_\eta * \rho(t,x) := \int_x^{x+\eta} \omega_\eta(y-x)\rho(t,y)dy, \qquad \eta > 0. \tag{2}$$

In (1), (2), we assume the following hypotheses:

$$(\mathbf{H}) \quad \begin{array}{ll} f \in \mathbf{C^1}(I;\mathbb{R}^+), & I = [a,b] \subseteq \mathbb{R}^+, \\ v \in \mathbf{C^2}(I;\mathbb{R}^+) \text{ s.t.} & v' \leq 0, \\ \omega_\eta \in \mathbf{C^1}([0,\eta];\mathbb{R}^+) \text{ s.t.} & \omega_\eta' \leq 0, \\ \int_0^\eta \omega_\eta(x)dx := J_0,\ \forall \eta > 0, & \lim_{\eta\to\infty} \omega_\eta(0) = 0. \end{array}$$

In this model, the function $\rho(t,x)$ represents the density of vehicles on the road, $\eta > 0$ is proportional to the look-ahead distance of drivers and v is the velocity function. Indeed, the aim is to describe the reaction of drivers who adapt their speed with respect to a weighted mean of the downstream traffic density. In particular, this class of equations includes the traffic flow models presented in [7, 20, 26, 30] for different choices of f and v, which are detailed below.

In [30], the authors introduce a traffic flow model based on Arrhenius stochastic microscopic dynamics. Using scaling and limit arguments, they obtain a macroscopic description of the microscopic dynamics leading to higher-order dispersive partial differential equations. The dynamics includes interactions with other vehicles ahead. The non-local traffic flow model that derives from this stochastic process is

$$\begin{cases} \partial_t \rho + \partial_x(\rho(1-\rho)\exp(-\omega_\eta * \rho)) = 0, \\ \rho(0,x) = \rho_0(x), \end{cases} \tag{3}$$

where the kernel acts only on the space variable x in this way

$$\omega_\eta * \rho(t,x) = \int_x^{x+\eta} \frac{J_0}{\eta} \rho(t,y) dy. \tag{4}$$

The kernel ω_η is an anisotropic short range inter-vehicle interaction potential, η is proportional to the look-ahead distance and J_0 is the interaction strength. Numerical solutions in [25] show that solutions of (3)–(4) allow for shock formations in finite time. In [26], the authors show that the finite time blowup of the solution of (3)–(4) must occur at the level of the first-order derivative and the L^p, $1 \le p \le \infty$, norms of the solution are finite close to the blow-up time. In this way, the results confirm the formation of shock waves for this non-local model.

While pedestrians are likely to react to the presence of people all around them, drivers mostly adapt their velocity to the downstream traffic, assigning a greater importance to closer vehicles. In [7, 20], the authors consider the following mass conservation equation for traffic flow with non-local velocity depending on a mean downstream density:

$$\partial_t \rho(t,x) + \partial_x \left(\rho(t,x) v \left(\int_x^{x+\eta} \rho(t,y) \omega_\eta(y-x) dy \right) \right) = 0,$$

for $t \in \mathbb{R}^+$ and $x \in \mathbb{R}$, $\eta > 0$. The kernel function $\omega_\eta \in \mathbf{C}^1([0,\eta];\mathbb{R}^+)$ is non-increasing, and the support η is proportional to the look-ahead visibility. The mean speed function v is continuous and decreasing. In [5], the authors also study the regularity results for the solutions of this non-local model proving Sobolev estimates and the convergence of approximate solutions solving a viscous non-local equation. The Blandin–Goatin model belongs to the class of equations considered in [21] and [9]. In [19], the authors propose a general non-local vehicular traffic flow model based on a mean downstream traffic velocity. This is the main difference with the models in [7, 9], where drivers adapt their velocity with respect to a mean downstream traffic density. In particular, the model in [19] allows to capture changes in the velocity function, and for this reason, it could be extended to junctions, as in [11].

2.1 Analytical Results for a Class of Scalar Non-local Traffic Flow Models

Adding an initial condition

$$\rho(0, x) = \rho_0(x), \qquad x \in \mathbb{R}, \tag{5}$$

with $\rho_0 \in \mathrm{BV}(\mathbb{R}; I)$, entropy weak solutions of the Cauchy problem (1), (5) are intended in the following sense [2, 6, 24].

Definition 1 A function $\rho \in (\mathbf{L^1} \cap \mathbf{L^\infty} \cap \mathrm{BV})(\mathbb{R}^+ \times \mathbb{R}; I)$ is an entropy weak solution of (1), (5), if

$$\begin{aligned}\int_0^{+\infty} \int_{\mathbb{R}} \{&|\rho - \kappa|\varphi_t + \operatorname{sgn}(\rho - \kappa)(f(\rho) - f(\kappa))v(\omega_\eta * \rho)\varphi_x \\ &- \operatorname{sgn}(\rho - \kappa) f(\kappa) v'(\omega_\eta * \rho)\partial_x(\omega_\eta * \rho)\varphi\} dx dt + \int_{\mathbb{R}} |\rho_0(x) - \kappa|\varphi(0, x) dx \geq 0\end{aligned} \tag{6}$$

for all $\varphi \in \mathbf{C^1_c}(\mathbb{R}^2; \mathbb{R}^+)$ and $\kappa \in \mathbb{R}$.

The following results are obtained in [9], approximating the problem (1), (5) through a Lax–Friedrichs-type numerical scheme and recovering $\mathbf{L^\infty}$ and BV estimates for the sequence of approximate solutions. Stability with respect to the initial data is obtained from the entropy condition using the doubling of variable technique.

Theorem 1 (Well-Posedness) *Let hypotheses* **(H)** *hold and* $\rho_0 \in \mathrm{BV}(\mathbb{R}; I)$. *Then the Cauchy problem* (1), (5) *admits a unique weak entropy solution* ρ^η *in the sense of Definition 1, such that*

$$\min_{\mathbb{R}}\{\rho_0\} \leq \rho^\eta(t, x) \leq \max_{\mathbb{R}}\{\rho_0\}, \quad \textit{for a.e. } x \in \mathbb{R},\ t > 0. \tag{7}$$

Moreover, for any $T > 0$ *and* $\tau > 0$, *the following estimates hold:*

$$TV(\rho^\eta(T, \cdot)) \leq e^{C(\omega_\eta)T}\, TV(\rho_0), \tag{8a}$$

$$\left\|\rho^\eta(T, \cdot) - \rho^\eta(T - \tau, \cdot)\right\|_{\mathbf{L^1}} \leq \tau e^{C(\omega_\eta)T} \left(\left\|f'\right\| \|v\| + J_0\|f\| \left\|v'\right\|\right) TV(\rho_0), \tag{8b}$$

with $C(\omega_\eta) := \omega_\eta(0) \left(\left\|v'\right\| \left(\left\|f'\right\| \|\rho_0\| + 2\|f\|\right) + \frac{7}{2} J_0 \|f\| \left\|v''\right\|\right)$.

Under more regular assumptions, an estimate of the dependence of the solution with respect to the kernel function, the speed and the initial datum can be proven, see [12]. Above, and in the sequel, we use the compact notation $\|\cdot\|$ for $\|\cdot\|_{\mathbf{L^\infty}}$.

Corollary 1 (Limit Model as $\eta \to +\infty$) *Let hypotheses* **(H)** *hold and* $\rho_0 \in \mathrm{BV}(\mathbb{R}; I)$. *As* $\eta \to \infty$, *the solution* ρ^η *of* (1), (5) *converges in the* $\mathbf{L^1_{loc}}$*-norm to the unique entropy weak solution of the classical Cauchy problem*

$$\begin{cases} \partial_t \rho + \partial_x (f(\rho) v(0)) = 0, & x \in \mathbb{R}, t > 0 \\ \rho(0, x) = \rho_0(x), & x \in \mathbb{R}. \end{cases} \tag{9}$$

In particular, we observe that $C(\omega_\eta) \to 0$ *in* (8a) *and* (8b), *allowing to recover the classical estimates.*

We observe that when the look-ahead distance $\eta \to \infty$, the non-local problem (1), (5) becomes a classical transport equation (9). Besides the mathematical implications, Corollary 1 may give information on connected autonomous vehicle flow characteristics. Indeed, large kernel supports could account for the information range between connected autonomous vehicles. On the other hand, when the visibility $\eta \to 0$, the non-local problem (1)–(5) is again reduced formally to a local problem, but the singular local limit is more challenging from the mathematical point of view. The derivation of this limit was initially conjectured in [2] for the one-dimensional scalar case motivated by numerical evidence, later corroborated in [7]. See [16, 17, 22, 33] for more details regarding the analytical derivation of the singular local limit for non-local conservation laws.

3 Non-Local Multi-Class Traffic Flow Models

In this section, we consider the following class of non-local systems of M conservation laws in one space dimension, introduced in [10]:

$$\partial_t \rho_i(t, x) + \partial_x (\rho_i(t, x) v_i((r * \omega_i)(t, x))) = 0, \qquad i = 1, \dots, M, \tag{10}$$

where

$$r(t, x) := \sum_{i=1}^{M} \rho_i(t, x), \quad v_i(\xi) := v_i^{\max} \psi(\xi), \tag{11}$$

$$(r * \omega_i)(t, x) := \int_x^{x+\eta_i} r(t, y) \omega_i(y - x) dy, \tag{12}$$

and we assume

(H1)
$\omega_i \in \mathbf{C^1}([0, \eta_i]; \mathbb{R}^+), \eta_i > 0$, s.t. $\omega_i' \leq 0$, $\int_0^{\eta_i} \omega_i(y) dy = J_i$.
$W_0 := \max_{i=1,\dots,M} \omega_i(0)$.
$v_i^{\max}$ are the maximal velocities $0 < v_1^{\max} \leq v_2^{\max} \leq \dots \leq v_M^{\max}$.
$\psi : \mathbb{R}^+ \to \mathbb{R}^+$ is a smooth non-increasing function s.t.
$\psi(0) = 1$, $\psi(r) = 0$ for $r \geq 1$.

For simplicity, we can consider the function $\psi(r) = \max\{1 - r, 0\}$. We couple (10) with an initial datum

$$\rho_i(0, x) = \rho_i^0(x), \qquad i = 1, \dots, M. \tag{13}$$

The model takes into account the distribution of heterogeneous drivers and vehicles characterized by their maximal speeds and look-ahead visibility in a traffic stream. It is a non-local generalization of the n-populations model for traffic flow described in [4], and it is a multi-class version of the one-dimensional scalar conservation law with non-local flux proposed in [7], where ρ_i is the density of vehicles belonging to the i-th class, η_i is proportional to the look-ahead distance and J_i is the interaction strength. The authors consider the following definition of weak solution because solutions to non-linear conservation laws are in general discontinuous, even if the initial datum is smooth.

Definition 2 A function $\boldsymbol{\rho} = (\rho_1, \dots, \rho_M) \in (\mathbf{L^1} \cap \mathbf{L^\infty})([0, T[\times\mathbb{R}; \mathbb{R}^M)$, $T > 0$, is a weak solution of (10), (13) if

$$\int_0^T \int_{-\infty}^{\infty} (\rho_i \partial_t \varphi + \rho_i v_i (r * \omega_i) \partial_x \varphi)(t, x) dx\, dt + \int_{-\infty}^{\infty} \rho_i^0(x) \varphi(0, x)\, dx = 0,$$

for all $\varphi \in \mathbf{C^1_c}(]-\infty, T[\times\mathbb{R}; \mathbb{R})$, $i = 1, \dots, M$.

The main result in [10] is the proof of existence of weak solutions to (10), (13), locally in time.

Theorem 2 *Let $\rho_i^0(x) \in (\mathrm{BV} \cap \mathbf{L}^\infty)(\mathbb{R}; \mathbb{R}^+)$, for $i = 1, \dots, M$, and assumption **(H1)** hold. Then the Cauchy problem* (10), (13) *admits a weak solution on* $[0, T[\times\mathbb{R}$, *for some $T > 0$ sufficiently small.*

4 An Upwind Numerical Scheme

For the numerical simulations in Sect. 5, we consider the following conservative scheme for the model (10)–(13) introduced in [19] in the scalar case and generalized in [10] to the multi-class case. First of all, we extend $\omega_i(x) = 0$ for $x > \eta_i$. For $j \in \mathbb{Z}$ and $n \in \mathbb{N}$, let $x_{j+1/2} = j\Delta x$ be the cell interfaces, $x_j = (j - 1/2)\Delta x$ the cells centres and $t^n = n\Delta t$ the time mesh. We approximate the initial datum ρ_i^0 for $i = 1, \dots, M$ and the kernel as follows:

$$\rho_{i,j}^0 = \frac{1}{\Delta x}\int_{x_{j-1/2}}^{x_{j+1/2}} \rho_i^0(x)\, dx, \qquad j \in \mathbb{Z}. \qquad \omega_i^k := \frac{1}{\Delta x}\int_{k\Delta x}^{(k+1)\Delta x} \omega_i(x)\, dx, \qquad k \in \mathbb{N},$$

so that $\Delta x \sum_{k=0}^{+\infty} \omega_i^k = \int_0^{\eta_i} \omega_i(x)\,dx = J_i$. We underline that the sum is finite since $\omega_i^k = 0$ for some $k \geq N_i$. Moreover, we set $r_{j+k}^n = \sum_{i=1}^{M} \rho_{i,j+k}^n$ for $k \in \mathbb{N}$ and

$$V_{i,j}^n := v_i^{\max} \psi \left(\Delta x \sum_{k=0}^{+\infty} \omega_i^k r_{j+k}^n \right), \qquad i = 1, \ldots, M, \quad j \in \mathbb{Z}. \tag{14}$$

We obtain the following upwind scheme:

$$\rho_{i,j}^{n+1} = \rho_{i,j}^n - \lambda \left(\rho_{i,j}^n V_{i,j+1}^n - \rho_{i,j-1}^n V_{i,j}^n \right), \tag{15}$$

where we have set $\lambda = \frac{\Delta t}{\Delta x}$.

5 Numerical Tests

In this section we show some numerical simulations of the model (10)–(13) for $M = 3$. The space step is $\Delta x = 0.001$. We use the numerical scheme in Sect. 4 to approximate the solution of (10)–(13).

5.1 Test 1: Impact of Connected Autonomous Trucks on a Circular Road

We study the impact of connected autonomous trucks on road traffic performances on a circular road. For this reason, we consider the space interval $[-1, 1]$ with periodic boundary conditions at $x = \pm 1$. In this case model (10)–(13) reads

$$\begin{cases} \partial_t \rho_1(t,x) + \partial_x \left(\rho_1(t,x) v_1^{\max} \psi((r * \omega_1)(t,x)) \right) = 0, \\ \partial_t \rho_2(t,x) + \partial_x \left(\rho_2(t,x) v_2^{\max} \psi((r * \omega_2)(t,x)) \right) = 0, \\ \partial_t \rho_3(t,x) + \partial_x \left(\rho_3(t,x) v_3^{\max} \psi((r * \omega_3)(t,x)) \right) = 0, \\ \rho_1(0,x) = \alpha\,(0.5 + 0.3 \sin(5\pi x)), \\ \rho_2(0,x) = \beta\,(0.5 + 0.3 \sin(5\pi x)), \\ \rho_2(0,x) = (1 - \alpha - \beta)\,(0.5 + 0.3 \sin(5\pi x)), \end{cases} \tag{16}$$

with

$$
\begin{aligned}
\omega_1(x) &= \frac{1}{\eta_1}, && \eta_1 = 0.3,\\
\omega_2(x) &= \frac{2}{\eta_2}\left(1-\frac{x}{\eta_2}\right), && \eta_2 = 0.05,\\
\omega_3(x) &= \frac{2}{\eta_3}\left(1-\frac{x}{\eta_3}\right), && \eta_3 = 0.1,\\
\psi(\xi) &= \max\{1-\xi, 0\}, && \xi \geq 0,\\
v_1^{max} &= 0.8, \quad v_2^{max} = 1.3, \quad v_3^{max} = 0.8.
\end{aligned}
$$

Above ρ_1 represents the density of autonomous trucks, ρ_2 the density of human-driven cars and ρ_3 the density of human-driven trucks. $\alpha \in [0, 1]$ is the penetration rate of autonomous trucks and $\beta \in [0, 1]$ is the penetration rate of human-driven cars. Autonomous trucks are modelled with a maximum velocity $v_1^{\max} = 0.8$ and a large kernel support $\eta_1 = 0.3$, because the interaction radius of connected autonomous vehicles is much greater than that one of human-driven cars. Moreover, we can assign a constant convolution kernel ω_1, since we assume that the degree of accuracy of information they have about traffic ahead is transmitted through wireless connections and does not depend on the distance. Figure 1 displays the traffic dynamics in the case $\alpha = 0.3$ and $\beta = 0.5$, and Fig. 2 displays the traffic dynamics in the case $\alpha = 0$ and $\beta = 0.5$ (no presence of autonomous trucks). We observe that oscillations are reduced if autonomous trucks are not present on the road, see Fig. 3.

5.2 *Test 2: Stretch of Straight Road with Autonomous Trucks*

In this test case, we propose a stretch of road with human-driven cars and trucks, and autonomous trucks. We consider the space domain $[-1, 1]$, imposing absorbing boundary conditions at the boundaries $x = \pm 1$. The initial conditions for Eq. (10) with $M = 3$ are the following:

$$
\begin{aligned}
&\rho_1(0, x) = \alpha_1 \chi_{[-0.6,-0.1]}(x), && \omega_1(x) = \frac{1}{\eta_1}, && \eta_1 = 0.5, && v_1^{max} = 0.8,\\
&\rho_1(0, x) = (1-\alpha_1)\chi_{[-0.6,-0.1]}(x), && \omega_2(x) = \frac{2}{\eta_2}\left(1-\frac{x}{\eta_2}\right), && \eta_2 = 0.1, && v_2^{max} = 0.8,\\
&\rho_3(0, x) = 0.5\chi_{[-0.9,-0.6]}(x), && \omega_3(x) = \frac{2}{\eta_3}\left(1-\frac{x}{\eta_3}\right), && \eta_3 = 0.05, && v_3^{max} = 1.3,\\
&\psi(\xi) = \max\{1-\xi, 0\}, \ \xi \geq 0.
\end{aligned}
$$

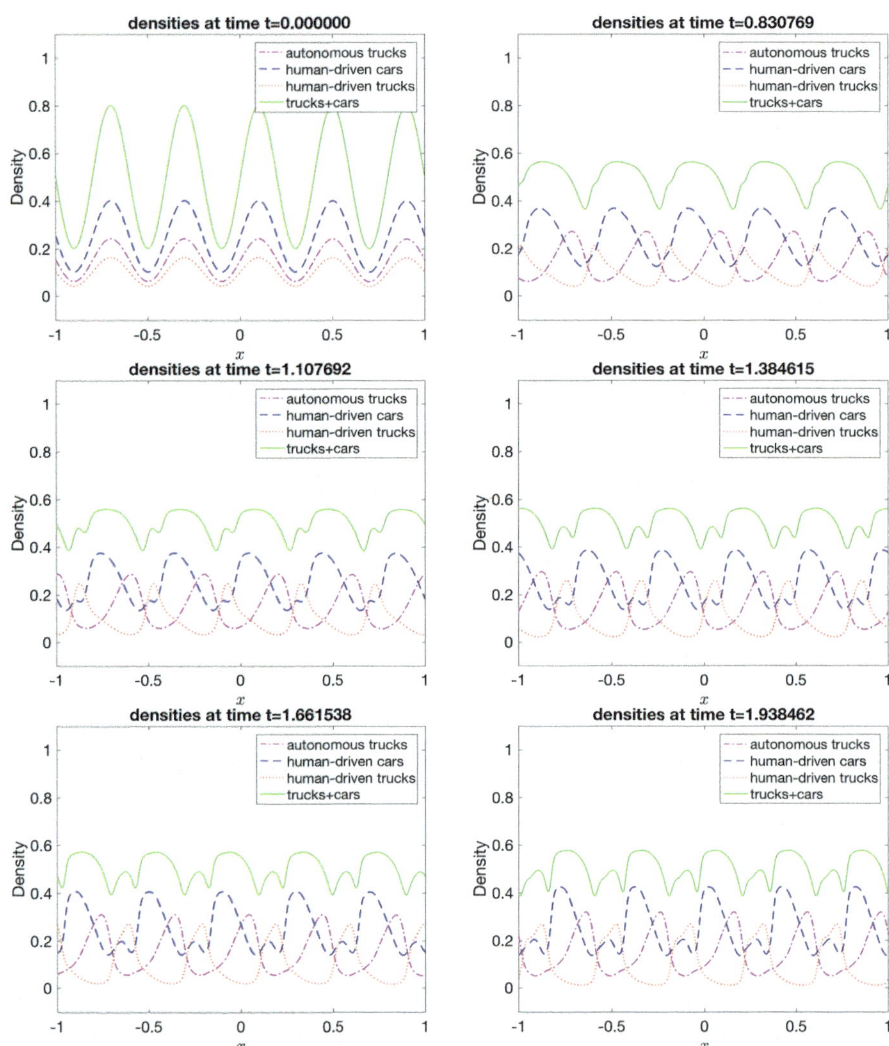

Fig. 1 Density profiles corresponding to the non-local problem (16) with $\alpha = 0.3$ and $\beta = 0.5$ at different times

Above ρ_1 represents the density of autonomous trucks, ρ_2 the density of human-driven trucks and ρ_3 the density of human-driven cars. $\alpha_1 \in [0, 1]$ is the penetration rate of autonomous trucks. Figure 4 displays the traffic dynamics in the case $\alpha_1 = 0.1$ and $\alpha_1 = 0.4$. In this case, the presence of autonomous trucks reduces the vehicle densities during the overtaking phase, see Fig. 4. The reason is in the fact that autonomous trucks have a very high look-ahead visibility η_1, and this enables them to adapt their velocity with respect to a weighted mean of the traffic density on a long stretch of road ahead.

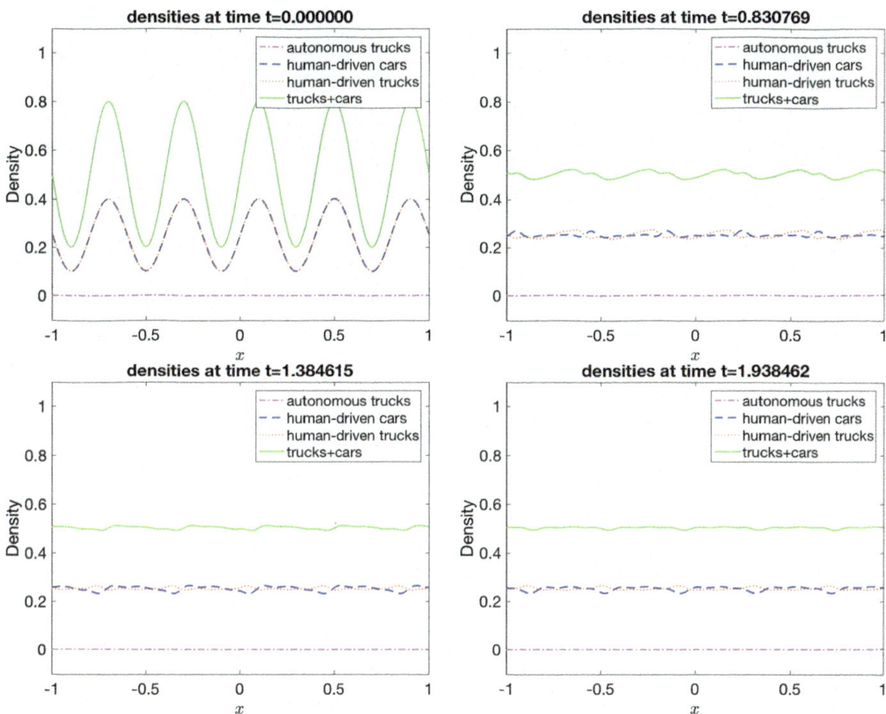

Fig. 2 Density profiles corresponding to the non-local problem (16) with $\alpha = 0$ and $\beta = 0.5$ at different times

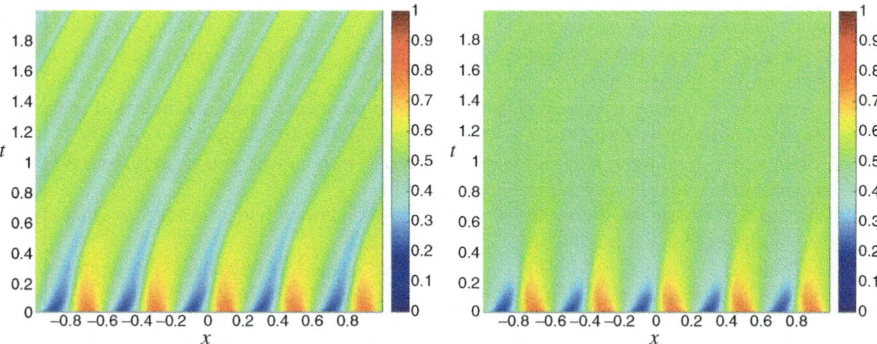

Fig. 3 (t, x)-plots of the total density $r(t, x) = \rho_1(t, x) + \rho_2(t, x) + \rho_3(t, x)$ computed with the upwind scheme (15), corresponding to different penetration rates of autonomous trucks: (**a**) mixed autonomous/human-driven traffic and (**b**) fully human-driven traffic

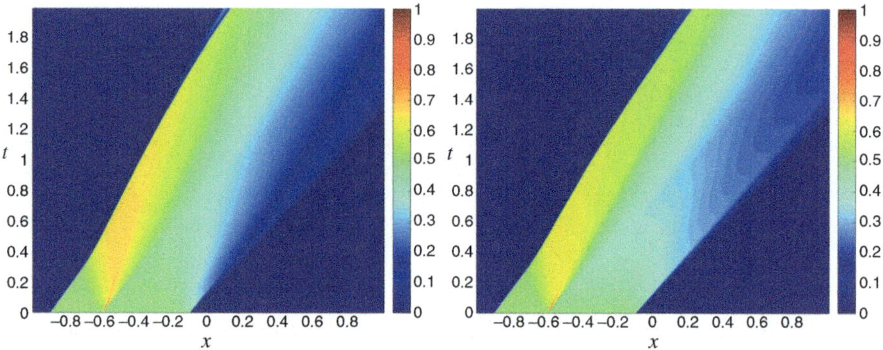

Fig. 4 (t, x)-plots of the total density $r(t, x) = \rho_1(t, x) + \rho_2(t, x) + \rho_3(t, x)$ computed with the upwind scheme (15), corresponding to different penetration rates of autonomous trucks: (**a**) $\alpha_1 = 0.1$ and (**b**) $\alpha_1 = 0.4$

Acknowledgments The author wishes to thank Paola Goatin for her suggestions and valuable discussions.

References

1. A. Aggarwal, R.M. Colombo, P. Goatin, Nonlocal systems of conservation laws in several space dimensions. SIAM J. Numer. Anal. **53**(2), 963–983 (2015)
2. P. Amorim, R. Colombo, A. Teixeira, On the numerical integration of scalar nonlocal conservation laws. ESAIM M2AN **49**(1), 19–37 (2015)
3. A. Aw, M. Rascle, Resurrection of "second order" models of traffic flow. SIAM J. Appl. Math. **60**(3), 916–938 (electronic) (2000)
4. S. Benzoni-Gavage, R.M. Colombo, An n-populations model for traffic flow. Eur. J. Appl. Math. **14**(5), 587–612 (2003)
5. F. Berthelin, P. Goatin, Regularity results for the solutions of a non-local model of traffic flow. Discrete Contin. Dyn. Syst. **39**(1078–0947 2019 6 3197), 3197 (2019)
6. F. Betancourt, R. Bürger, K.H. Karlsen, E.M. Tory, On nonlocal conservation laws modelling sedimentation. Nonlinearity **24**(3), 855–885 (2011)
7. S. Blandin, P. Goatin, Well-posedness of a conservation law with non-local flux arising in traffic flow modeling. Numer. Math. **132**(2), 217–241 (2016)
8. C. Chalons, P. Goatin, L.M. Villada, High-order numerical schemes for one-dimensional nonlocal conservation laws. SIAM J. Sci. Comput. **40**(1), A288–A305 (2018)
9. F.A. Chiarello, P. Goatin, Global entropy weak solutions for general non-local traffic flow models with anisotropic kernel. ESAIM Math. Model. Numer. Anal. **52**(1), 163–180 (2018)
10. F.A. Chiarello, P. Goatin, Non-local multi-class traffic flow models. Netw. Heterog. Media **14**(2), 371–387 (2019)
11. F.A. Chiarello, J. Friedrich, P. Goatin, S. Göűttlich, O. Kolb, A non-local traffic flow model for 1-to-1 junctions. Eur. J. of Appl. Math. **31**(6), 1029–1049 (2020)
12. F.A. Chiarello, P. Goatin, E. Rossi, Stability estimates for non-local scalar conservation laws. Nonlinear Anal. Real World Appl. **45**, 668–687 (2019)
13. F.A. Chiarello, P. Goatin, L.M. Villada, High-order Finite Volume WENO schemes for non-local multi-class traffic flow models, in *Hyperbolic Problems: Theory, Numerics, Applications* (2019)

14. F.A. Chiarello, P. Goatin, L.M. Villada, Lagrangian-antidiffusive remap schemes for non-local multi-class traffic flow models. Comput. Appl. Math. **39**(2), 60 (2020)
15. R.M. Colombo, M. Herty, M. Mercier, Control of the continuity equation with a non local flow. ESAIM Control Optim. Calc. Var. **17**(2), 353–379 (2011)
16. M. Colombo, G. Crippa, M. Graffe, L.V. Spinolo, Recent results on the singular local limit for nonlocal conservation laws (2019). arXiv:1902.06970
17. M. Colombo, G. Crippa, L.V. Spinolo. On the singular local limit for conservation laws with nonlocal fluxes. Arch. Rational Mech. Anal. **233**(3), 1131–1167 (2019)
18. J. Friedrich, O. Kolb, Maximum principle satisfying CWENO schemes for nonlocal conservation laws. SIAM J. Sci. Comput. **41**(2), A973–A988 (2019)
19. J. Friedrich, O. Kolb, S. Göttlich, A Godunov type scheme for a class of LWR traffic flow models with non-local flux. Netw. Heterog. Media **13**(4), 531–547 (2018)
20. P. Goatin, S. Scialanga, Well-posedness and finite volume approximations of the LWR traffic flow model with non-local velocity. Netw. Heterog. Media **11**(1), 107–121 (2016)
21. A. Keimer, L. Pflug, Existence, uniqueness and regularity results on nonlocal balance laws. J. Differ. Equ. **263**(7), 4023–4069 (2017)
22. A. Keimer, L. Pflug, On approximation of local conservation laws by nonlocal conservation laws. J. Appl. Math. Anal. Appl. **475**(2), 1927–1955 (2019)
23. A. Keimer, L. Pflug, M. Spinola, Existence, uniqueness and regularity of multi-dimensional nonlocal balance laws with damping. J. Math. Anal. Appl. **466**(1), 18–55 (2018)
24. S.N. Kružkov, First order quasilinear equations with several independent variables. Mat. Sb. (N.S.) **81**(123), 228–255 (1970)
25. A. Kurganov, A. Polizzi, Non-oscillatory central schemes for a traffic flow model with Arrhenius look-ahead dynamics. Netw. Heterog. Media **4**(3), 431–451 (2009)
26. D. Li, T. Li, Shock formation in a traffic flow model with Arrhenius look-ahead dynamics. Netw. Heterog. Media **6**(4), 681–694 (2011)
27. M.J. Lighthill, G.B. Whitham, On kinematic waves. II. A theory of traffic flow on long crowded roads. Proc. R. Soc. London. Ser. A. **229**, 317–345 (1955)
28. H. Payne, *Models of Freeway Traffic and Control* (Simulation Councils, Incorporated, 1971)
29. P.I. Richards, Shock waves on the highway. Oper. Res. **4**, 42–51 (1956)
30. A. Sopasakis, M.A. Katsoulakis, Stochastic modeling and simulation of traffic flow: asymmetric single exclusion process with Arrhenius look-ahead dynamics. SIAM J. Appl. Math. **66**(3), 921–944 (electronic) (2006)
31. G. Whitham, *Linear and Nonlinear Waves* (Pure and applied mathematics. Wiley, 1974)
32. H. Zhang, A non-equilibrium traffic model devoid of gas-like behavior. Transp. Res. B Methodol. **36**(3), 275–290 (2002)
33. K. Zumbrun, On a nonlocal dispersive equation modeling particle suspensions. Quart. Appl. Math. **57**(3), 573–600 (1999)

Printed by Printforce, the Netherlands